Principles of

CONTAMINANT HYDROGEOLOGY

Second Edition

Christopher M. Palmer

Consulting Hydrogeologist
San Jose, California

CRC Press
Taylor & Francis Group
Boca Raton London New York

CRC Press is an imprint of the
Taylor & Francis Group, an **informa** business

CRC Press
Taylor & Francis Group
6000 Broken Sound Parkway NW, Suite 300
Boca Raton, FL 33487-2742

© 1996 by Taylor & Francis Group, LLC
CRC Press is an imprint of Taylor & Francis Group, an Informa business

First issued in paperback 2019

No claim to original U.S. Government works

ISBN-13: 978-0-367-44851-6 (pbk)
ISBN-13: 978-1-56670-169-3 (hbk)

Visit the Taylor & Francis Web site at
http://www.taylorandfrancis.com

and the CRC Press Web site at
http://www.crcpress.com

Library of Congress Card Number 96-4967

Library of Congress Cataloging-in-Publication Data

Palmer, Christopher M.
    Principles of contaminant hydrogeology / by Christopher M. Palmer.
    -- 2nd ed.
        p. cm.
    Includes bibliographical references and index.
    ISBN 1-56670-169-4
    1. Groundwater--Pollution. 2. Hydrogeology. I. Title.
TD426.P35 1996
628.1′68--dc20

                                                                    96-4967
                                                                    CIP

Cover Design: Shayna Murry

PREFACE

This book is intended to be an introduction for newcomers, and a refresher for professionals, on basic principles of consulting contaminant hydrogeology. Much of the material presented here has grown out of an introductory groundwater monitoring course taught by the author for the University of California Extension, Santa Cruz, and other seminars. Many students taking the course were eager to become involved in groundwater issues and cleanup, yet few realized what was involved in applied subsurface investigation. Also, many were unaware of investigating the unsaturated and saturated zones and their importance in complying with regulations for site cleanup. Hence, this text presents and reviews problem-solving approaches commonly used by consulting geologists and hydrogeologists to meet regulatory guidance and meet obligations to clients.

An accurate subsurface contamination problem assessment within the context of applicable regulations for industrial or municipal clients is usually the consulting hydrogeologist's goal. Accuracy of investigation data will have profound effects on defining the contaminant extent and site remediation effectiveness and cost. Consequently, all investigative efforts should be expended toward understanding the subsurface processes that affect contaminant movement, remediation effectiveness, and site closure.

An emphasis is placed on the basic geologic nature of contaminant hydrogeologic investigations. The hydrogeology may then be understood within the geologic framework — that is, aquifers, aquitards, water quality, porosity, permeability, and so on. Davis (1987) points out rightly that the field of hydrogeology occurs within geologic science, and not in engineering or other fields. The site geology is too often simplified or not adequately understood, which can bring grief to a consultant. This is not to say that other disciplines of engineering, chemistry, and laboratory analysis are not needed or should be minimized. A contamination investigation often is a multidisciplinary effort. However, individuals who practice hydrogeology should have a broad academic training in geology and extensive field experience in performing subsurface data collection and interpretation studies. Hence, the need to properly perform sampling and testing techniques for hydrogeologic meaning at each step is critical. In this way, interpretation of contaminant movement becomes more comprehensible.

Although the book introduces general aspects of an investigation, the need to understand site geology, stratigraphy, and aquifer properties is the main goal. The location of contaminants and success of remediation depend on this. Inadequately defined subsurface geology often results in a lack of subsurface understanding that compromises knowledge of groundwater and contaminant movement. Typically, projects may have limited budgets and resources with which to study a contamination problem. There will never be enough money and exploratory boreholes for absolute answers to all subsurface questions due to the variability of the geologic materials.

The quantity and quality of subsurface site hydrogeology collected by the consultant form the basis for successful problem definition. Judgments of data adequacy must be made in every subsurface investigation, since each site investigation is unique. Both the data and the models will be extrapolated across the site to cover areas where information is lacking. Therefore, the subsurface information should be the best attainable given the time, money, and regulatory constraints so that data extrapolation will not be questioned as unreasonable. The ultimate test of both the hydrogeologic data adaquacy and extrapolation occurs when the remedial system is installed and cleanup performance is monitored.

Various interrelated aspects to investigation execution such as field logistics, drilling techniques, sampling protocols, general concepts of contaminant movement, and remediation are presented. Professionals involved in this work should keep abreast of the literature, new equipment and analysis techniques, groundwater and related textbooks, and regulatory rules changes, which can occur rapidly. The novice hydrogeologic consultant or regulator overseeing contaminant hydrogeologic projects often needs to weave different threads of the study so that the report information makes sense. While technical project objectives must be achieved, the legal, financial, and ethical objectives in consulting are as important. This book is neither a "magic bullet" with all the answers nor a "cookbook" approach for success in hydrogeologic consulting. Rather, it seeks to introduce neophyte hydrogeologists — as well as chemists, engineers, regulators, and lawyers — to basic information gathering so that sufficient subsurface data is available to build the basic geologic, hydrogeologic, and contaminant models.

Consultants must also deal with project budgets, client interaction, and regulatory requirements. When a client signs a contract with the consultant, it is usually to perform a specific scope of work, typically the investigation to solve the client's problem. However, this means that the geologist or hydrogeologist must understand the applicable regulations and the legal implications therein. The project budget has been proposed to do a certain work scope and the amount of information collected will be a direct function of the available money. A vexing problem that can arise is, how much information is collected for the time and money expended? The investigation being performed may be reconnaissance or preliminary, and formal "comprehensive reports" may require several investigation phases before the problem is adaquately defined.

Often the consulting hydrogeologist produces technical reports that will be read by a nontechnical audience. Information contained in reports should present the immediate problem, in clear terms supported by the site data, even when it may be "bad news." The consultant should realize that his or her work may be critically reviewed years later by other consultants or regulating agencies. The information gathered in the subsurface investigation must be defensible and current with what others in the field are or were doing, the so-called "state of the art."

It is hoped that this book will help both new and experienced consultants, regulators, lawyers, and any generally interested individuals by outlining consulting approaches to hydrogeologic projects.

AUTHOR

Christopher M. Palmer is a practicing geologist and hydrogeologist in San Jose, California. He holds the BA and MA in geology and has practiced as a consultant since 1979. He is a registered geologist in California, Arkansas, Florida, and Pennsylvania, and is a registered hydrogeologist and certified engineering geologist in California. Mr. Palmer has been an instructor for the University of California Extension, Santa Cruz, since 1988 in the Hazardous Materials Certificate Program, and has presented other instructional seminars for the Geological Society of America and the U.S. Environmental Protection Angency. Mr. Palmer's interests are applied geology and hydrogeology, contaminant transport, and site assessment and cleanup.

ACKNOWLEDGMENTS

The second edition of this book has been extensively rewritten, I thank Jeffrey Peterson and Dr. Jerold Behnke, who helped on the first edition as well as the second. I especially want to thank the individuals who lent assistance to me for this second edition of *Principles of Contaminant Hydrogeology*. Steve Beck, Doug Young, and Jim Rubin were the reviewers for this edition; their insights and critiques have made the book stronger in geology, hydrogeology, and remediation engineering, respectively. Dr. Jan Turk, Dave Guthridge, and John Elliott gave help by providing time to discuss various aspects of hydrogeology, contracting remediation approach and regulatory interaction, and their viewpoints from their working experience. Perspective and reality insights by Jerry Tappa and Larry Pavlak and my family are much appreciated. I thank all the individuals and publishers who kindly gave permission to reproduce material included in the text and figures. Finally, thanks are due to all consultants, geologists, and hydrogeologists with whom I have worked and discussed the various subjects contained in the text. Much text material includes workday approaches used by consultants in everyday applied contaminant hydrogeology problem-solving.

DISCLAIMER

The material and approaches presented in the text of this book reflect the views or opinions of the author. Mention of products, agencies, persons, or equipment in the text of this book is not meant to be an endorsement.

Malo mori suam foe oami.

Par sit fortuna labori.

TABLE OF CONTENTS

Principles of
CONTAMINANT HYDROGEOLOGY
Second Edition

CHAPTER 1

Geologic Frameworks for Contaminant Hydrogeologic Investigations

INTRODUCTION

Subsurface investigations into the presence and extent of contaminated groundwater are primarily geological investigations. The site subsurface geology forms the physical framework through which groundwater and contaminants move. Understanding the site geology provides the fundamental basis for understanding site hydrogeology and defining contaminant movement. This information is of paramount importance when preparing models for groundwater flow, contaminant transport and fate, and the site remediation plan. Indeed, the first model erected in the study is the geologic model, which is overlain by the hydrogeologic model, contaminant distribution model, and so forth. Geologic environments will vary depending on where the geologist is working, but the basic hydrogeologic questions of depth to groundwater, aquifer contacts, and so on need to be answered.

The investigations also integrate aspects of soil engineering, applied chemistry, and environmental engineering disciplines. Although some surface geophysical techniques may aid subsurface exploration, site information is typically required to be collected from exploratory boreholes for soil, rock, sediment, soil vapor, and groundwater sampling. In fact, some of these requirements have been codified in portions of state and federal regulations and government guidance documents as the industry has matured in the past years.

Site investigations are conducted to gather the following information regardless of the site size or potential contaminant problem in the unsaturated (vadose) zone and saturated zone. Current regulations and cleanup are directed to both the groundwater and overlying soil and sediment that could have a continuing impact on groundwater quality. Thus, the investigation should address geology and operational processes of the vadose zone and aquifer. This will yield information on the general site geology, hydrogeology, aquifer geometry, groundwater occurrence and flow direction, gradient, physical and chemical testing programs, and aquifer pump testing. In this way, the consulting geologist and hydrogeologist can meet the primary goal of identifying and tracking contaminants for site cleanup.

GROUNDWATER OCCURRENCE AND GEOLOGY OF AQUIFERS

Groundwater occurs in subsurface rock and strata called aquifers, which comprise porous and permeable material (see Figure 1). Aquifers may be composed of alluvium, sedimentary rocks, and fractured crystalline or metamorphic rocks. The aquifers can be bounded by relatively "impermeable" bodies called aquitards, which do not readily transmit water. The discussions throughout most of the rest of the text will treat aquifers in a stratigraphic sense — that is, products of sediment deposition with sand and gravel as aquifers and silt and clay as the aquitard (or units of interest grouped by composition as aquifers and aquitards). Fractured rock aquifers will be discussed separately below. Aquifers and aquitards will be conceptually presented as tabular bodies that, although it is a strong generalization, will illustrate subsurface exploration and movement of water and contaminants. When case histories deal with specific hydrogeologic conditions, the text will so state.

A BRIEF REVIEW OF UNSATURATED AND SATURATED
GEOLOGIC ENVIRONMENTS

Groundwater occurs in almost every type of geology. Therefore, an exhaustive review of all geologic and hydrogeologic environments is beyond the scope of this book. Groundwater occurrence has been described in the standard texts and the reader is referred to those for detailed discussions (see, for example, Davis and DeWeist, 1966; Fetter, 1988; Freeze and Cherry, 1979; Heath, 1982). A review of the unsaturated zone will be presented first, followed by a review of the general concepts of groundwater occurrence and flow.

Unsaturated (Vadose) Zone

The vadose zone overlies the saturated zone, or, for our purpose, any aquifer. Typically there is a contaminant release; the contaminants must pass through this region to get to the aquifer. Although this region is not saturated (that is, having all available pore space filled with fluid), areas may be locally saturated, while elsewhere fluid moves in response to tension and capillary forces. The vadose zone is also typically where volatile contaminant gases or vapors will arise and move from contaminant releases (see Figure 2).

The vadose zone is geologically a very heterogeneous region. Typically soil formation occurs at the surface and soils may be buried sequentially in fluvial and alluvial depositional processes. Unsaturated zones may vary in thickness

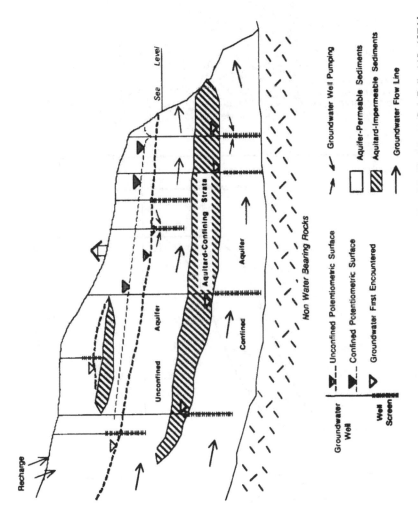

Figure 1 Hydrologic cycle with unconfined and confined aquifers (Modified from *Cal. Bull.*, 118, 1974.)

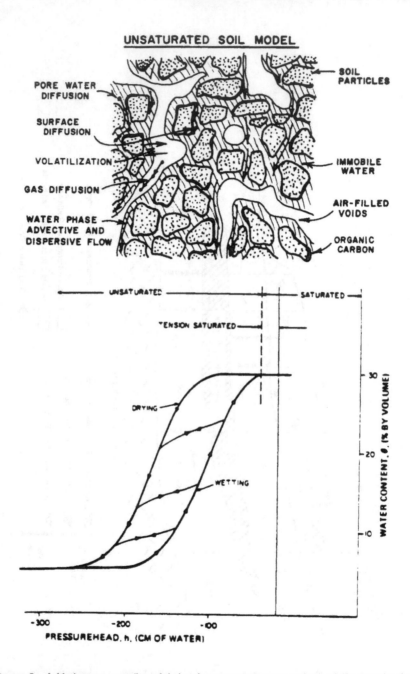

Figure 2   A. Vadose zone soil model showing mass transport and retardation mechanisms. (From Gierke, Hutzler and Crittenden, 1985. With permission.) B. Relationship between soil moisture, matric suction, and hydraulic conductivity. (From Freeze and Cherry, 1979 in EPA, 1993. With permission.)

from several feet to several hundred feet. Very thick vadose zones typically occur in the alluvial basins of the western U. S. The term "soil" has become a catch-all to describe anything that is not a rock, and may cause a somewhat geologically misleading interpretation. For example, the site of interest may be soil-covered fractured rock, which contains soil, colluvium, and rock. A series of soils (or paleosols) may be buried sequentially, thus resulting in a crudely horizontally bedded deposit many feet thick, below the organic rich surface soil and soil profile. Further, because most contamination problems occur in an urbanized setting, land development such as digging basements, pipelines, or foundations may have altered the subsurface. The range of particle sizes, bedding, presence of buried structures, and so on may have a profound effect upon soil vapor and liquid contaminant movement (Morrison, 1989). Accurately ascertaining the geologic makeup of the vadose zone is vital to understanding the migration pathways toward the groundwater. A substantial investigative effort is required to locate the contaminant and understand the local geologic implications.

Unsaturated fluid movement may be affected by one or more processes depending on local conditions. Four processes which affect movement are hysteresis, macropore flow, capillary movement, and saturated (or Darcian) flow.

Hysteresis

The unsaturated region is not devoid of moisture, but what moisture does occur adheres to soil particles and grains. Flow that occurs between grains (pore throats) moves through micropores. A general equation for unsaturated flow (Fetter, 1988) is

$$phi = psi(\phi) + Z, \text{ where}$$

phi - total potential, unsaturated flow
$psi(\phi)$ - moisture potential, measured as suction
Z - elevation head

Fluid movement is governed by fluctuations in local pressure gradients. The suction (usually measured in centibars per cubic centimeter) is the effect of a negative pressure head of soil and water on the unsaturated conductivity (Hillel, 1980; Morrison, 1989; see Figure 2). As the soil becomes saturated, the soil pressure or suction declines toward zero as the available porosity becomes totally filled with water. As the soil drains, suction increases until the remaining water is held in tension on or between soil particles. As moisture increases or decreases in the soil, the suction changes. This would also be the case for contaminants migrating into and through the soil. For contaminant studies, the most important difference between saturated and unsaturated flow is that the unsaturated hydraulic conductivity (K) is not a constant for the different soil moisture contents (Morrison, 1989).

Macropore Flow

This occurs when surface water enters a crack or macropore and literally flows under gravity down into the crack. The vertical movement continues until the crack fills and local saturated conditions arise migrating from the crack (see

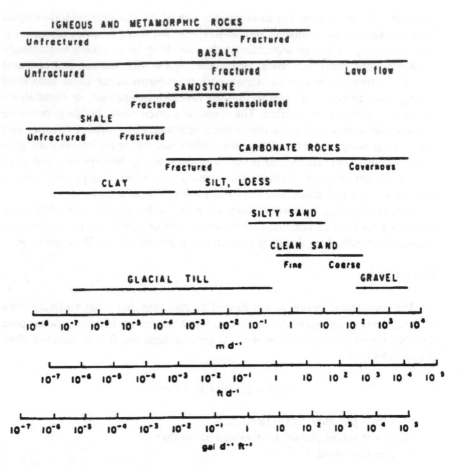

Figure 3 Hydraulic conductivity of selected consolidated and unconsolidated aquifers. (From Heath, 1982. With permission.)

Figure 3). Openings and channels are common in vadose soil and buried sediment resulting from roots, worms, burrowing animals, or man-made conduits and foundations. Desiccation cracks in clay may be open to depths of several feet, or even tens of feet, and could be infilled with granular sediment. When the clay is remoisturized, the sand in the crack may allow enhanced flow. This could allow rapid and deep penetrations of fluid through an otherwise "impermeable" layer with the obvious potential movement toward the aquifer. Lateral saturation from the vertical flow would occur at a slower rate into the soil around the macropore (Morrison, 1989; U.S. EPA, 1987). The implication that soil logging techniques must allow for recognition of these structures in the site investigation is obvious.

Capillary Movement

Capillarity occurs due to available porosity immediately above the saturated zone as a response to water surface tension allowing vertical movement against gravity. The height of the capillary "fringe" above the saturated area is dependent on grain size. Hence, the height of capillarity into a clay or silt is higher than that of a sand or gravel. The pores of the clay and silt are much smaller and movement by tension occurs (Hillel, 1980).

Saturated Flow

This occurs where sufficient water collects on an impermeable layer to saturate the porosity. This results in a "perched" groundwater lens above the regionally recognized aquifer. These perched lenses may be small or very extensive, and form local "miniaquifers." In this case, Darcian flow can occur and could be used to model the movement of fluid. If large quantities of water or contaminants are present, movement could be initially lateral and horizontal under unsaturated conditions, then vertical under saturated conditions.

Saturated Geologic Environments — Aquifers

This brief review of geologic environments that may be encountered in consulting hydrogeology may be classified broadly as igneous-metamorphic, sedimentary, and alluvial-filled basins. These three general classes will be used throughout the book as examples. Groundwater occurrence in these discussions is somewhat similar to the DRASTIC models developed by the U.S. Environmental Protection Agency for use in rating geology for contaminant investigations (Heath, 1984; Aller, et al., 1987a). The following summary is not an attempt to modify DRASTIC or other established hydrogeologic convention; the reader is referred to those references from which the following discussion draws. The intention is to introduce general groundwater occurrence in terms of porosity and permeability for different terrains (for an excellent review of this material, see Back et al., 1988).

The location and identity of drinking water aquifers are basic to contaminant hydrogeology since the goal is to protect that resource. Consulting work demands proper recognition of geologic materials and environments so that the appropriate investigation approach is selected to collect the proper and relevant data. Obviously, individual sites need to be evaluated for their specific hydrogeologic characteristics in all cases. Geologic materials vary in different rock types; Figure 3 presents diagrams and typical ranges of porosity and permeability of rocks and sediment (Heath, 1982; Morris and Johnson, 1967). Figure 4 shows idealized geologic units and groundwater flow.

Igneous–Metamorphic Rock

Crystalline igneous or metamorphic rocks, and some very well lithified or slightly metamorphosed sedimentary rocks, form these terrains. Usually these rocks are "dry" in a large-quantity water resource sense, yield little water to wells, and are considered impermeable. Groundwater flow is typically through joints, fractures, structural discontinuities, or tectonic fracturing, with very little porosity

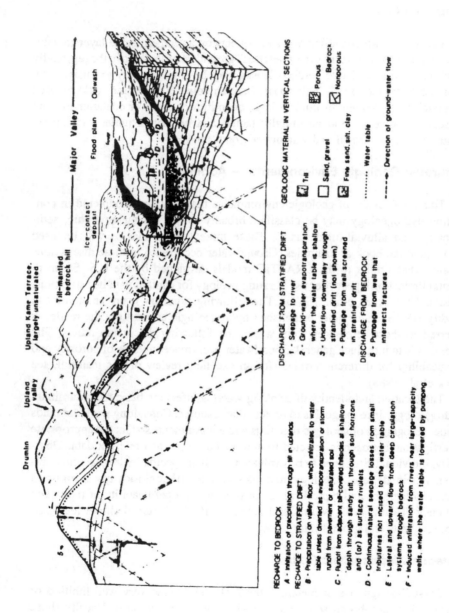

Figure 4 Idealized distribution of geologic units and groundwater flow. Not to scale. Major valleys actually occupy 5–30% of large basins. (From Randall, 1988. With permission.)

between minerals or metasediment grains. Where fractures are dense and recharge is abundant, locally high water yields may occur. In igneous and metamorphic terrains, fractures tend to be common in the upper 300 ft, and will contain and yield water. Overlying alluvial accumulations or weathered zones of rock add to the available porosity and water reservoir. While some fracture systems can be deep and extensive, typically there are fewer fractures and they close with depth and groundwater yield decreases accordingly. Although laminar flow assumptions may not be valid in fracture flow, it may not be a major hindrance to describing flow and transport in fracture systems (Schmelling and Ross, 1989).

Sedimentary Rock

Sedimentary rock contain lithified and semilithified sediments, as well as chemical sediments such as limestone. Porosity and permeability in sedimentary rocks tend to be greater than in igneous or metamorphic rocks. Groundwater resource occurrence is typically more abundant in sandstone and limestone formations and limited in siltstone and shaley formations. Water flow is usually around grains through pores or voids, although fracture flow may occur and could be a predominant flow path in well-lithified or cemented rocks. The "conventional" model of layered aquifer and aquitard systems can occur in these terrains, where tabular aquifer stratigraphy is usually envisioned.

Carbonate rock flow paths range from flow through fractures and around grains to flow through caverns. Flow situations may become somewhat unpredictable because the solution void creation and infilling, caves and sinkhole formation, presence of breccias and formation metamorphism, and internal stratification may affect water velocity and direction. The recharge by streams, other aquifers, and heavy precipitation events may flow directly into voids and exit through other voids further "downgradient," and may resurface as springs or streams. Tracers may be used to track water flow in carbonates. Carbonate rock hydrology is not discussed in this book; White (1988) presents a good general review of this and other topics.

INTRODUCTION TO ALLUVIAL-FILLED BASINS AND SEDIMENTARY DEPOSITIONAL ENVIRONMENTS

Alluvial-filled basins are depressions into which alluvial sediment is deposited by fluvial processes. These depositional processes may include alluvial fan, rivers, floodplains, glacial outwash, and lakes. These deposits can contain large usable groundwater resources in intermountain areas, especially in the western U. S. For example, the San Joaquin Valley (California) is underlain by thick sediment sections containing an enormous unconfined and confined aquifer system with tremendous groundwater reserves. Similar alluvial basins in Arizona and Nevada have been developed for the water needs of Phoenix and Las Vegas. Other alluvial basins may be quite small — for example, perched valleys within mountains with sufficient water for small cities or only individual homes. Groundwater occurrence is similar to that of sedimentary rock terrains; sand strata tend to form aquifers, and clay and silt strata form the aquitards in layered systems. Since large urbanized areas tend to occur on these relatively flat basin surfaces, the developed groundwater is threatened by more numerous contamination sources. For a more detailed breakdown of groundwater regions in the U. S., see Heath (1984).

Strata laid down in alluvial basins occur in certain depositional environments that dictate the geometry and occurrence of sand and clay bodies (Reineck and Singh, 1986; see Figures 5 and 6). The percentage of gravel, sand, silt, and clay, and the mixtures of these textures will profoundly affect their hydraulic conductivity. Sediment may be deposited in strata that are laterally continuous, or discontinuous over feet, yards, or miles, depending on the environment. Internal stratification causes sediment texture gradations and interbedding within the overall sand and clay bodies. Site geology and environment portrayal usually go hand in hand on the cross section with each report (Vishner, 1965). Glacial sedimentation may result in clayey lake sediments, channel sand and gravel, outwash alluvium, or a combination of these depending on the location. However, some prediction of the location of sedimentary strata is possible if the type of depositional environment is recognized and the investigator has some knowledge of the internal aspects of that environment. Unless some of this knowledge is used, the textural and lateral correlation aquifer and aquitard strata can be very difficult to interpret, making contaminant location equally difficult to predict underground.

Table 1 presents typical characteristics of active depositional environments in order to illustrate the distribution of sediment in these environments. Once the characteristics are recognized, then an idea of the position, lateral extent, and sediment texture can be estimated. This allows one to begin to conceptualize the aquifer strata and the possible layered relationship of the aquitards and aquifers. This can be a powerful tool in understanding the location of groundwater, assisting in recognition of possible contaminant pathways, and making estimates of water yield and directions of contaminant movement.

Alluvial Fan

The alluvial fan environment occurs at the base of the mountain ranges where rivers and streams draining the range flow onto relatively flat valley floors. Active tectonic activity causes the basins to subside and hence, a thick accumulation of sediment can occur at the mountains' base and into the valley floors. These deposits can be coarse grained where large particles are dumped as the streams exit the mountain range. As the streams flow downhill, the available energy required to carry sediment decreases, leaving boulder-sized particles but continuing to carry fine gravel, sand, silt, and clay. Occasionally mixtures of gravel, sand, silt, and clay move as one (debris flows). Although limited sorting may occur in distributary streams cutting through the fan, the stratigraphy is very crude with nonsorted, in that the sediment is not sorted but "dumped" in a basin at the foot of the mountains.

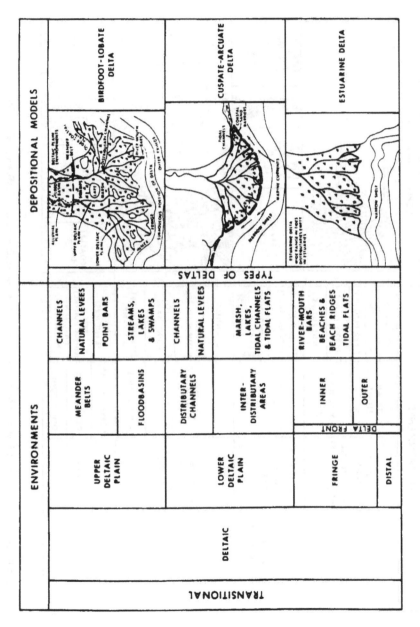

Figure 5   Deltaic depositional models of clastic sedimentation. (LeBlanc, 1971. With permission.)

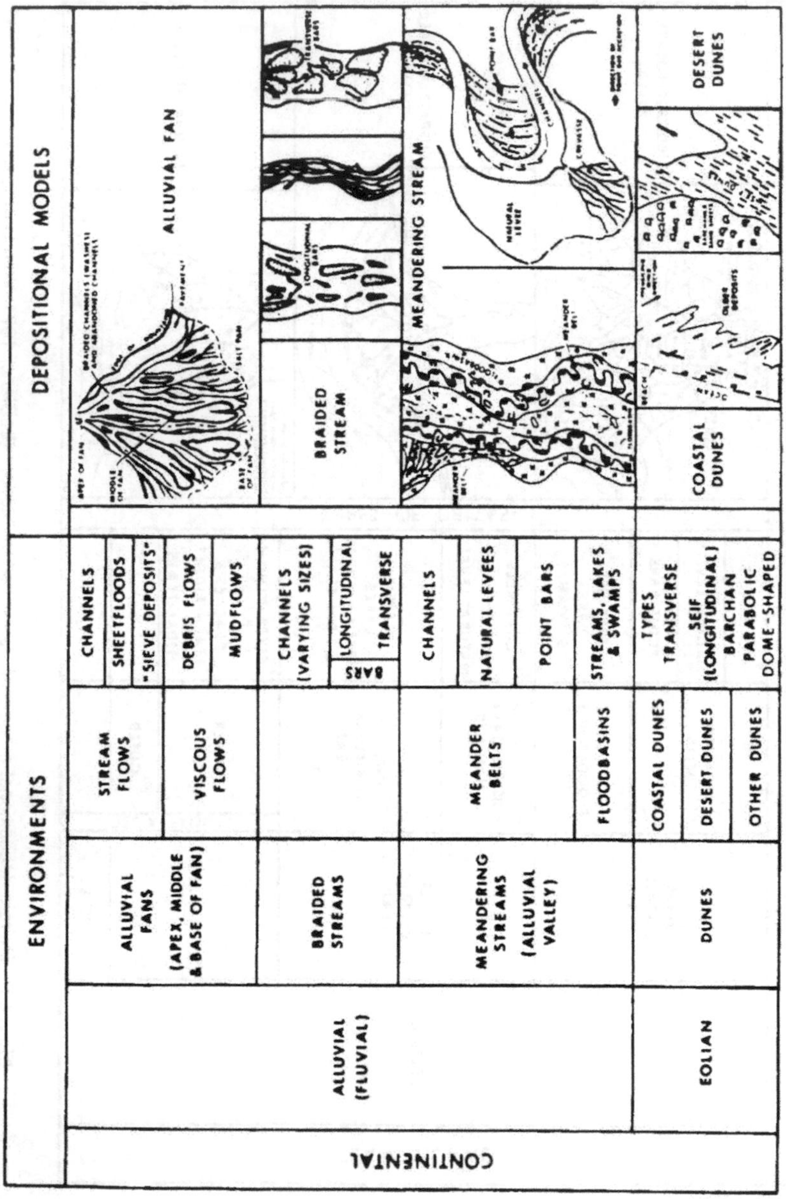

Figure 6 Alluvial (fluvial) and Eolian depositional models of clastic sedimentation. (LeBlanc, 1971. With permission.)

Table 1   Generalized Characteristics of Five Depositional Environments

Type of environment	Sediment type	"Typical" stratigraphy	Vertically cyclic deposits
Alluvial fan	Texture gravel through clay; one active channel, many abandoned channels.	Random, finer grained down-fan; sand/gravel strata weakly defined.	None; gravel abundant, sand in channels, all strata laterally and vertically discontinuous.
Braided river	Sand/coarse gravel, little silt or clay; numerous stream channels	Laterally extensive sand; interbedded gravel beds sand	Sand/gravel strata defined and stacked vertically, silt/clay strata thin and discontinuous
Meandering river (MR)	Gravel, sand silt and clay; one large developed channel	Strata laterally extensive, coarse gravel strata usually confined to stream channel deposits	Gravel/sand and silt/clay strata repeated vertically, both can be laterally continuous
MR/floodplain	Predominantly clay, sand occurs as laminae, sand and fine gravel in channels; large MR channel	Clayey with thin sand strata, rare fine gravel and sand channels	Clay strata laterally continuous; sand/fine gravel pipelike channels contained in clay
Glacial	Boulders and gravel through clay; varies depending on location and glacial history	Can be complex and random; outwash deposits sand/gravel till sand/clay; silt loess strata	None to cyclic deposits on glacial history; may have braided river strata; strata may be laterally continuous in till, outwash, and loess

Braided River

The braided river environment occurs downslope of the alluvial fan, although the two will be gradational. The braided river occupies large channels in the alluvial valley. The stream has energy to carry and sort a full sediment load of gravel, sand, silt, and clay, but the stream has sufficient energy to carry the silt and clay downstream. When the river flow has maximized channel capacity, the gravel is pushed into bars while the sand continues to flow around the gravel bars. When water flow falls to a certain level, the gravel bars halt in channel bends (due to less energy), while the sand is deposited around and on top of the gravel until the next high flow (storms, etc.). Sand strata are laterally and vertically extensive, while gravel strata are laterally and vertically discontinuous. When the river flow is very low, slack water deposits may allow silts and some clay to be deposited. These fine-grained deposits tend to be thin and not laterally extensive.

Meandering River and Floodplain

The meandering river occurs where the sediment deposition over time are large and river gradient is low. For example, the Mississippi River Valley is underlain by huge sediment accumulations (thousands to tens of thousands of feet thick). The river follows a sinuous path due to gradient, creating a looping net of channels. As it flows, fine gravel and coarse sand move in the deepest channel incision while lateral bank erosion occurs on its loops. This action lays down a cyclic deposit of gravel overlain by sand in the channel. The sand builds up in levee deposits, in response to episodic flooding. As the channel cuts into the circuitous loops, it eventually intersects the downstream portion of the channel, resulting in a "cut-off" commonly called an oxbow lake. The lake then fills in with silt and clay carried to the lake in times of flood, filling it. As this process occurs, organic material can be deposited as well, as these tend to be vegetation-rich environments. Where the river gradient is very flat, peat deposits can be formed in the delta. (Contrast this with the energy of the aforementioned two environments.)

The floodplain and channel deposits are interrelated. When the river floods, the silt and clay (suspended load) will move out of the channel and onto the floodplain. When the flood ebbs, the flood waters become slack, and the silt and clay settle out. Occasionally, the channel levee breaks and water flows onto these lower areas with the same effect. Because the sediments will compact and the basin will subside, silt and clay strata will collect and form laterally extensive deposits. Over time, they may cover channels, forming deposits where the sand and very fine gravel sediment is "contained" in the clay. The clay and silt strata contain fine-grained sand beds of laminae, but are very thin and not regionally laterally extensive. Channels may again loop back on themselves, cutting and filling previous deposits, creating a very complicated stratigraphy.

Glacial Deposits

Glacial deposits cover the northern portions of the U. S. to varying thicknesses, and in mountain valleys. Inasmuch as several glacial advances have occurred in the recent geologic past, older deposits may be overlain by others, or moved and redeposited. Stephenson et al. (1988) describe the following types of glacial deposits:

> Till: sediment deposited directly by the glacier without significant resorting with generally poor aquifer potential;
> Glaciofluvial: sediments deposited by streams associated with the glacier, which can contain sorted sand and gravel, with good to excellent aquifer potential;
> Glaciolacustrine: lake sediments deposited from glacial activity, with variable aquifer potential;
> Loess: silt deposits associated with glacial activity and landscapes — although they may be deposited by nonglacial processes, with variable aquifer potential (see Figure 4).

Stratigraphy can be highly variable and units may or may not be laterally and vertically continuous. Low permeability units may be fractured, which can enhance the ability of water to flow in them. Recharge into till and loess units may be through fractures developed in voids, weathering, and biologic processes. Where glacial outwash deposits are present, they may contain abundant "clean" sand and gravel with water yield potential. Jopling and McDonald (1975) present a review of glacial depositional processes.

A BRIEF REVIEW OF GROUNDWATER MOVEMENT

The following is a very brief review of groundwater movement and some of the general relationships of water flow commonly used in consulting work. This discussion borrows from the following groundwater references, but is not a substitute for referring to these texts (Domenico and Schwartz, 1990; Driscoll, 1986; Fetter, 1988; Freeze and Cherry, 1979; Heath, 1983) or for completing rigorous university groundwater or hydrogeologic classes, and groundwater mathematics instruction therein.

Subsurface hydrogeology of any area is typically divided into the vadose zone and saturated zone. Each zone comprises sediment or rock, which possesses a total porosity (e. g., open space). When the volume of open space is divided by the total volume and the result is multiplied by 100 (n) in percent. Permeability (k) can be generally described in qualitative terms as the ease with which fluid can move through a porous rock or sediment and is measured by the rate of flow in suitable units. Hydraulic conductivity (K) is the capacity of a porous medium to transmit water, and is used in consulting where it refers to the water-transmitting characteristic of material in quantitative terms. Hydraulic conductivity values may range over several orders of magnitude for sediment and rocks, and usually exhibit a range within the material itself from place to place (Heath, 1982). As a general rule, the horizontal hydraulic conductivity is greater than the vertical conductivity. If the hydraulic conductivity is essentially the same in all directions, then the aquifer is isotropic; if it varies in different directions, it is anisotropic. It is very rare to find natural deposits that have truly isotropic and homogeneous aquifer conditions.

Groundwater flows toward a direction of decreasing total head (sum of the elevation head and the point where pressure head is measured). Groundwater moves under a hydraulic gradient (i), which is the change in head per unit of distance in a given direction (usually calculated between wells as the difference in height divided by the length of incline or distance between wells). The relationship used to describe groundwater flow is Darcy's law, $Q = KAi$, where discharge (Q) equals hydraulic conductivity times cross-sectional area times gradient (see Figure 7).

Transmissivity (T) is the capacity of an aquifer to transmit water through a unit thickness of the aquifer where $T = Kb$, where b is the aquifer saturated thickness. The storage coefficient (S) is a dimensionless measure of water released from storage divided by the unit area times the unit head change (see Figures 7 and 8).

Groundwater flow can be mapped by using equipotential lines, which are lines connecting points of equal head, that shows the potentiometric surface for that aquifer. The lines of equal pressure are commonly called groundwater contour lines, plotted at a known elevation for the date they were measured. This represents a plane of equal pressure relative to atmospheric in the aquifer as it flows from areas of recharge to areas of discharge. In an unconfined aquifer, the potentiometric surface corresponds to the groundwater occurrence in that forma-

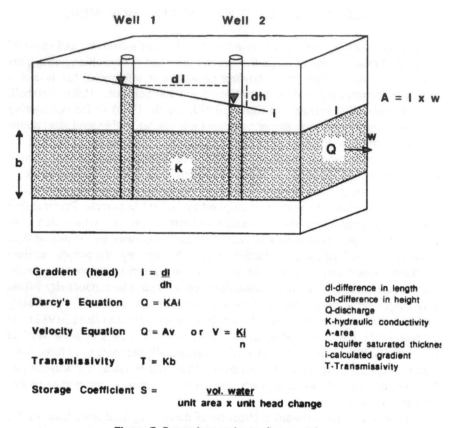

Figure 7 General groundwater flow equations.

tion. A confined aquifer is confined by an overlying stratum and the potentiometric surface is an imaginary plane measured in wells that penetrate the confining bed. The water levels rise in the wells above the confining bed to some elevation where the pressure becomes equal to atmospheric (see Figure 9).

Groundwater flow can be diagrammed with flow nets once the equipotential lines have been drawn. These flow nets are constructed by drawing flow lines perpendicular to the equipotential lines, and represent idealized paths by water particles as they move in the aquifer. The flow nets may easily diagram any aquifer flow and the presence of lateral or upward gradient. The water flow may change direction and gradient with changes in recharge or discharge. The gradient often displays seasonal or regional pumping influence changes, and the assumption of permanent flow directions and gradients can be grossly misleading during the investigation.

Aquifers are usually conceived as tabular bodies that are homogeneous, isotropic with distinct contacts or boundaries for ease in conceptual and mathematical presentation, but numerous exceptions to these assumptions exist in the field. These assumptions are made in order to quantify aquifer characteristics and describe the flow. The geology of the aquifer is highly varied and will exert local

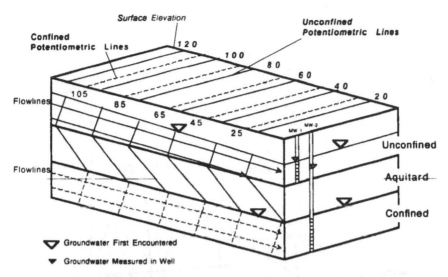

Figure 8 Potentiometric surfaces in unconfined (water table) and confined aquifers. Cutaway view of two aquifers and their potentiometric surfaces measured to an elevation datum on the surface. Wells MW-1 and MW-2 penetrate the aquifers. Water table corresponds to the unconfined potentiometric surface in MW-1; pressure surface is potentiometric surface in the confined aquifer in MW-2. (Modified from EPA, 1987.)

influences that can have obvious effects on the assumptions made to simplify the aquifer model. Data analysis must take the site-specific geology into account when applying the appropriate mathematics to analyze the groundwater flow.

Perched Groundwater

Perched groundwater conditions arise when water collects upon an impermeable stratum above what would be considered the regional aquifer. Perched aquifers are usually limited in areal extent, and the water quality and quantity are highly variable. Recharge may be natural or man-made (sewers, agricultural runoff, flood basins) or a combination of both. Over a sufficiently long time, the groundwater in the perched lens would ultimately continue to migrate downward as regional aquifer recharge. Occasionally, these aquifers may be large enough to be locally recognized as water producers.

Perched aquifers may form in numerous ways, as illustrated in the following examples. Clay strata contained within a sand strata may form an impermeable layer upon which the water collects. Alluvium or colluvium (alluvial sediment that collects on hillsides) overlying relatively impermeable bedrock may collect water at or above the sediment-rock contact. Lastly, water may collect in interconnected fractures or joints in bedrock, but vertical movement is restricted by aperture width, infilling, or closing of fractures at depth. Hence, the water tends to "flow" downhill in the fractures and is perched in the otherwise impermeable bedrock. In many instances, the perched water is usually the first impacted by descending contaminants.

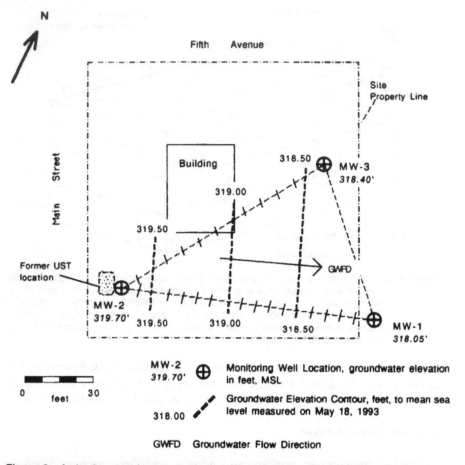

Figure 9 A simple groundwater contour map. The project map shows the wells with the contours measured for the date shown. The contours are drawn by measuring the distance between two wells and dividing that distance by the fall in elevation between two wells, and repeating for each well. The difference between the wells elevation can be divided into roughly equal lengths and the contours interpolated given the calculation and site observations. A contour map is an interpretative map with the groundwater flow line is drawn perpendicular to the contour line.

Groundwater Movement in Fractured Rock

As discussed above, the movement of groundwater in fractured rock (igneous, metamorphic, and some carbonate rock, and some clay tills or clayey deposits) is typically governed by the occurrence, aperture, and density of fractures in the otherwise solid rock. Although the same mathematical approaches are used in fractured rock, they are usually modified and assumptions for flow modeling must be made. Intergranular flow is usually negligible in the sense of the previous discussion in that the contribution of water is minimal, if any. The fractures will control the quantity and the flow direction.

The fracture opening or aperture allows the water to move in the rock; however, the fractures may be constricted or infilled, which would inhibit or deflect flow, or they may be open and highly interconnected, allowing more flow. Fractures may occur along or subparallel to bedding planes, depositional contacts, tectonic and structural discontinuities, solution features, dessication, joints, internal rock contacts, and exfoliation; they could be enhanced in areas of mining, landsliding, roadcuts, and deep weathering profiles. Although fractures may tend to close near the surface (within roughly 300 feet of the surface) as a general rule, they can be open and yield water hundreds of feet below the surface. Fracture mapping can be very useful in finding patterns of intersecting fractures and mapping their relation to springs, known wells, and creeks or streams.

Water development potential may be small when compared to alluvial basin fills or some carbonate aquifers. Fractures may "daylight" directly to the surface or through soil or alluvial covers so macropore flow can become very important in recharge. If fractures connect more porous sediments or a zone of interconnected fractures, flow direction may be preferential due to the occurrence of openings. Fracture location using surface mapping and downhole logging and testing is vital to evaluating groundwater flow paths because the contaminants will be assumed to follow the same paths.

Discharge can be into streams or creeks and in mountain valley alluvial fill, or intermountain valley fill (see Figure 4). Mapping groundwater flow and contaminant routes is challenging and may range from easy to very difficult, depending on the local situation. Modeling of these fracture flow paths as well as pump-testing methods must be suitable and applicable, or modified for use for fractured situations. For general reviews of fractured flow, see Hitchon and Bachu (1988), Bear et al. (1993), and Gustafson and Krasny (1994).

CONCEPTUAL APPROACH TO CONTAMINANT HYDROGEOLOGIC INVESTIGATIONS

A hydrogeologic investigation for contamination will involve a subsurface study so that site-specific conditions are understood. The geologic and hydrogeologic approach is to determine the extent of site contamination for a cleanup effort. The site geology will form the physical framework through which the groundwater and contaminants flow. Thus, the geologic model should initially describe hydrogeologic constraints and may arguably be the most important modeling step. The accuracy of the geologic study is directly influenced by the experience of the investigator, subsurface drilling and mapping, sampling, and analysis. Boreholes are expensive to drill, and local access and legal restraints may limit placement of boreholes. Consequently, conceptual layout of the study is needed so that the information is as complete and accurate as possible for each borehole drilled given the time and budget constraints.

Consulting hydrogeologists may conduct contamination investigations on sites that are only hundreds of thousands of square feet (consider that an acre is 43,560 square feet). Thus, one is looking at very small areas in detail, whereas classical U. S. Geological Survey (USGS) water supply studies commonly encompass tens to hundreds of square miles. A contaminant investigation in an urban area will be evaluating shallow saturated zones (the uppermost water-bearing zone or strata), which have probably not been previously studied or used for water resources. Therefore, all the information will usually have to be generated from the site geological study. Very small features such as strata several inches or feet thick and potential contaminant pathways (e. g., animal burrows or root holes) can become highly significant. The so-called "aquifers" of interest may be only a few feet thick, and those strata may occur as discontinuous beds or lenses with complex sedimentologic gradations. The field approach must collect not only the same types of information as one would for an areally large geologic investigation, but also the detailed and small-scale features.

As much information as possible should be reviewed prior to initiating field activity, including all the usual regional sources of geologic and hydrogeologic information in libraries or universities. The USGS is an excellent source of information contained in water supply papers, local and regional studies and water quality and geochemical data in the nationwide STORET computer database. The EPA and individual states produce guidance manuals for data collection, local geologic work, and contaminant-related studies. The state and local agency files are particularly good for very localized information, which has been or is being filed as public record in response to contamination events and has grown over the past several years. Although research into the movement and fate of contaminants is limited, it is growing and the investigator should review the literature periodically to keep abreast of new trends.

Other information sources are available and should be used as the need arises. Local city and county planning and engineering offices keep geotechnical engineering reports on file that contain subsurface boring logs, depth to groundwater, soil types, well details, and so on, all of which is public record and available for review. Local water districts or water management and flood control agencies maintain records of well locations, surface stream flow, groundwater pumping history, areawide hydrogeology, and water-use history. In response to hazardous materials management laws and regulations, local fire and emergency response agencies may have records of materials stored at industrial locations, building permits for special containment structures, and subsurface storage tanks. This allows site history of potential contaminant sources to be compiled and is especially useful when contaminant plumes cross political and property boundaries.

Finally, the consultant may wish to discuss the site problem with experts in chemistry, regulations, law, and engineering. Contamination problems are complex and almost always require interdisciplinary approaches. Furthermore, the regulations are becoming more complex and definitive as to the types and completeness of the information required. The consultant must perform within budget and time restraints that can be very challenging, and it is rare that the consultant gets all the time, money, and field data desired. The more information one has at the beginning of the project, the more complete the problem definition and the ability to formulate solutions.

WHAT IS IMPORTANT IN INITIAL SITE AQUIFER INVESTIGATION APPROACH?

You have been retained to begin a site investigation. Where do you start and what do you do? First, site location, history, and access are required so that you can get an initial idea of what had gone on there. While it seems obvious, many times the site location, physical layout, or geography of the site is unclear. For example, the site may have been an industrial facility in the past, then was dismantled, and now is is an open field. You should ask for the client's entire site files and for access to people knowledgeable about past site history. Next, you might go to the city planning office to review any geotechnical reports or building permits, and it helps to get a site map or plan with the original or modified plant layout. Then you should always go to the site for a first-hand look. There is nothing like walking the site to get a feel for the geography surrounding industrial activities. Also, just because it is a flat lot does not preclude remnants of the former operation like foundation outline, subsurface pipe vent lines, patched asphalt, and so on. It also allows you to envision possible borehole locations and logistical constraints, such as whether the drilling rig will move on pavements or sink in a muddy field.

Up until now, you have not done much geologic work, and rightfully so. When consulting, the investigator is often responsible for managing the project from the start, including many preliminary details that have very little to do with geologic technical expertise. Now that you have an idea of what the place is like, you can estimate the number of boreholes and the sampling depths. This will be expanded on later, but a review of the regional geologic information and the local planning or engineering office or water district allows an initial guess of the depth to groundwater, flow direction, and soil and sediment types you will encounter. While some information exists regarding site chemical usage, an outline for sample depths and laboratory analytical program must be finalized.

You know that the geology controls the site groundwater flow and pollutant movement, so you can apply your knowledge of vadose and saturated zone processes and use them to your advantage. Regional geologic information should tell you the type of rocks and sediment, and the depositional environment, or presence of fractured rocks. This can be used to estimate the initial presence of sand and clay strata, their three-dimensional continuity, their lateral extent, and their hydraulic characteristics. The site-specific data must of course come from the subsurface study, but now you have a general idea of the geology, hydrogeology, site history, and possible contaminants. These data address the early portion of the site study, and help you anticipate the regulatory issues and potential contaminant extent so the client has an idea of the problem facing him.

AN EXAMPLE OF "CHANGEABLE" AQUIFERS

A spill site adjacent to a river has been monitored with several shallow and deep monitoring wells for months. The initial aquifer interpretation appears to be thin and discontinuous sandy clay that grades to a thick, sandy aquifer at depth. This is interpreted to be a river floodplain deposit underlain by a large channel deposit laid down by previous channel meander. Shallow wells yield water slowly and deep wells yield water relatively easily.

The observations in monitoring wells are puzzling because gasoline product and groundwater containing dissolved contaminants are periodically present in the shallow wells. The deeper wells yield "clean" water and never contain floating product. When the water table surface falls below the the shallow wells, floating product does not enter the deeper wells as one would surmise, assuming that the immiscible product would sink as water levels decline. Overall, the dissolved plume does not seem to migrate in the direction suggested by the deeper wells, but remains stationary.

The site is near a large river that rises and falls in response to precipitation and floods as irrigation dams release water. As the river rises, the shallow wells recharge and release water slowly, as expected given the depositional environment. When the river falls, the water levels in the shallow wells decline, as does the floating product and dissolved contaminants, but the deeper wells are still "clean." A cross section of the site stratigraphy shows that the upper clay "caps" the lower sand, which is in hydraulic connection with the river through a transitional sandy clay (see Figure 10). When river flow increases and water levels rise, it crosses the sandy clay-sand contact and water appears in the shallow wells (see Figure 11). When the river falls, the shallow wells display a vertical gradient to the lower sand unit. The contaminants in the shallow wells begin to move vertically down toward the sand. However, the river flow is sufficient to prevent this floating contaminant from penetrating across the contact, so the deeper wells are not affected. The groundwater movement of interest in the upper aquifer is really vertical rather than horizontal.

Here the upper clay "aquifer" appears aquitard-like, but transmits some water when the river rises. Because contaminants have affected groundwater at the site, the first occurrence of groundwater is the "uppermost aquifer" in the regulatory view. Both "high" and "low" permeability sediments are involved, and the aquifer is actually both units combined. While apparently two potentiometric surfaces are observed, the upper wells respond more slowly due to the clayey sediment. Gasoline is more or less "smeared" in the upper clayey unit that with rising water levels, allows product to reenter the well as both floating and dissolved product in groundwater. The alluvial deposit at this site controls the water movement in the wells, and in a small area appears as anomalous water levels and contaminant movement. The clay yields less water and contaminant as water levels decline. The point is that the entire site must be viewed within the geologic framework where thin sand layers contained in clay in the upper unit may behave somewhat

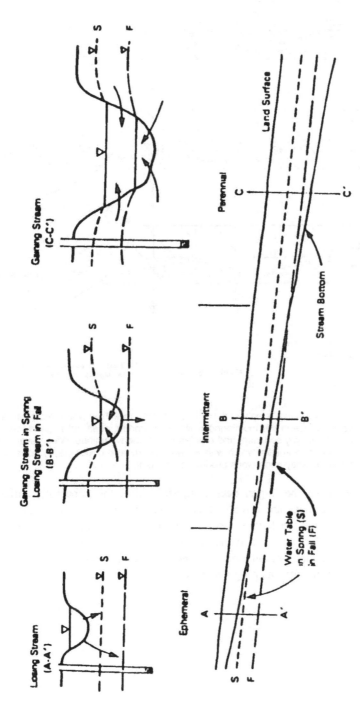

Figure 10 Relationship between water table level and seasonal stream water level (From EPA, 1990. With permission.)

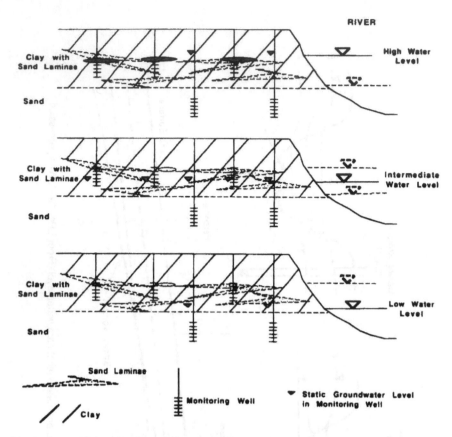

Figure 11   Floating product is only observed during high river flow when the water levels
cause a rise in the monitoring wells. The product does not migrate vertically
due the clayey sediment and resides in the sandy laminae. Water levels move
through one aquifer which with a lower sandy and upper clayey zone cause
the apparently anomolous product movement.

independently from the larger, deeper aquifer. Also, the apparently different
potentiometric surfaces are actually related because the wells were placed to
observe the slower water level response. Finally, the contaminant is essentially
"hung up" in the clayey upper unit and migration is limited by the upward gradient
so as to prevent its moving through the sandy clay and into the larger aquifer, a
fortunate occurrence for this property owner.

CHAPTER 2

Contaminant Pathways — Subsurface Investigation and Monitoring Approach

INTRODUCTION

When a contaminant is introduced at the ground surface, it must migrate through the unsaturated zone toward the aquifer. Hence, contaminants will move through soil, sediment, fractured rock, manmade conduits, or other pathways on their way to the saturated zone. Investigations need to ascertain the horizontal and vertical extent of contaminants in both unsaturated and saturated zones. Additionally, the existing soil and groundwater quality must be quantified for background comparisons. It is important to establish preexisting contamination location or natural background concentrations that may exceed a regulated standard.

Needs and Goals of the Hydrogeologic Investigation Approach

The ultimate goal of this type of hydrogeologic study is to ascertain the extent of the contamination so that an approach for remediation can be developed. The needs and goals of the investigation and the site conditions must be considered to effectively position exploratory borings and monitoring wells. Today this can include a regulator's review of the work plan and approval of it prior to starting field work. Site-specific knowledge of aquifer stratigraphy, contaminant type, and suspected pathway will greatly influence well design and investigation procedures' success. Over the past several years, this approach has become somewhat standardized to the government regulations and guidance documents for site subsurface investigations. The field rationale for borehole and well placement to define contamination should then address the following;

- define site geology and stratigraphy
- define hydrogeology (water occurrence gradient and flow)
- collect soil samples for vertical and horizontal vadose contaminants delineation
- collect groundwater samples to define the dissolved plume extent
- allow flexibility in data collection for overall site coverage

It is very important to remember that the hydrogeologist consultant wants to gain the maximum information for the time and money spent in the field that is required to answer the regulatory questions and solve the client's problem. Groundwater quality changes as it moves through the contaminated area and influences downgradient quality. Otherwise, clean sites may become contaminated from the upgradient plume sources. The client should not clean up a problem if he did not create it simply because it moved onto his site. Since a property or political boundary may be crossed by the water, the problem grows in terms of responsible parties and their potential liabilities. Parties often debate the cause and who "owns" the problem; however, the longer the plume moves and disperses, the larger the contaminated area becomes and the more comprehensive, complex, and expensive the geologic investigation and ultimate remediation.

Types and Sources of Contaminants

Contaminants are "unnatural" substances introduced in the subsurface that degrade natural or existing groundwater quality. Obviously a complete list of contaminants is beyond this discussion; however, the following list of contaminant groups is presented. Classifying contaminants into groups may assist the investigator since industrial or other processes may utilize certain materials that give rise to working contaminant "suites" (as solvents and certain heavy metals for microchip manufacture, or petroleum hydrocarbons at a refinery or service station). Such contaminant groups may include inorganic "heavy" metals (trace elements), and other organic materials including hydrocarbon fuels, greases and oils, industrial solvents and chemicals, herbicides and pesticides, coolants, explosives, and agricultural chemicals and wastes (U.S. EPA,1986).

The sources of these materials are widespread and numerous. Again, a classification grouping may tie together the types of contaminants of interest for the investigation (Miller, 1980). Such a classification grouping follows (see Table 1 and Figure 1):

- Municipal — sewers, sanitary landfills, disposal wells, military bases
- Industrial — manufacture, subsurface tanks, pipelines, mines, oil fields
- Agricultural — fertilizers, pesticides, wastes, irrigation return flow
- Other — transport spills, material stockpiles, septic tanks, testing labs

Once these materials migrate from a source, they must move along some route or pathway into soil or rock and toward groundwater (see Figure 1). It should be noted that "natural" contaminants exist in soil and groundwater and the site study must differentiate these from the contamination problem. The existence of trace elements, biologic, and even naturally occurring petroleum compounds may appear to be contaminants when they are indigenous to the site. Also, industrialized areas may show a background of contaminants if those materials had a widespread use. Hence, the trace concentration level may or may not initially suggest a problem unless supported by site-specific data and analysis.

Table 1 Typical Sources of Groundwater Contamination by Land Use Category

Category	Contaminant source	
Agriculture	Animal burial areas	Irrigation sites
	Animal feedlots	Manure-spreading areas/pits
	Fertilizer storage/use	Pesticide storage/use
Commercial	Airports	Jewelry/metal plating
	Auto repair shops	Laundromats
	Boat yards	Medical institutions
	Construction areas	Paint shops
	Car washes	Photography establishments
	Cemeteries	Railroad tracks and yards
	Dry cleaners	Research laboratories
	Gas stations	Scrap and junkyards
	Golf courses	Storage tanks
Industrial	Asphalt plants	Petroleum production/storage
	Chemical manufacture/storage	Pipelines
	Electronics manufacture	Septage lagoons and sludge sites
	Electroplaters	Storage tanks
	Foundries/metal fabricators	Toxic and hazardous spills
	Machine/metalworking shops	Wells (operating/abandoned)
	Mining and mine drainage	Wood-preserving facilities
Residential	Fuel oil	Septic systems, cesspools
	Furniture stripping/refinishing	Sewer lines
	Household hazardous products	Swimming pools (chemical storage)
	Household lawns	
Other	Hazardous waste landfills	Recycling/reduction facilities
	Municipal incinerators	Road deicing operations
	Municipal landfills	Road maintenance depots
	Municipal sewer lines	Stormwater drains/basins
	Open burning sites	Transfer stations

From U.S. EPA, 1993b. With permission.

CONTAMINANT PROPERTIES AFFECTING TRANSPORT

The chemistry of the contaminants will affect their transport and fate. Moore and Ramamoorthy (1984) list these properties in two groups: physicochemical properties such as solubility, vapor pressure, partition coefficient, sorption-desorption, volatilization; and chemical transformations such as oxidation-reduction behavior, hydrolysis, halogenation-dehalogenation, and photochemical breakdown. These processes will be discussed briefly below, and this following review borrows heavily from the work of Moore and Ramamoorthy (1984) and Lewis (1993). Other information charts and reference books can provide chemical data for the contaminant or material of interest.

Physiochemical Processes

Solubility is the degree and ease to which the chemical or compound will dissolve in water. Solubility determines the concentration present in water and if the contaminant will interact with other chemicals and to the extent that they become molecularly or ionically dispersed in the solvent to form a true solution. The precise determination of solubility remains elusive for many contaminant compounds, and some of the aqueous solubility values are only estimates. Many environmentally sensitive compounds have very low water solubilities. Miscibility

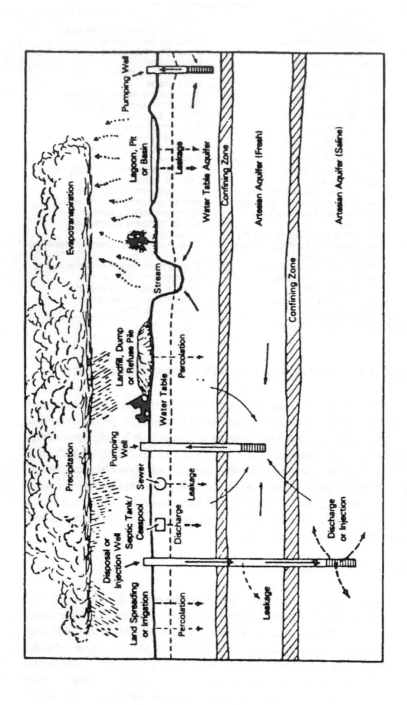

Figure 1   Conceptual input of contaminants into the subsurface and possible migration pathways. (From U.S. EPA, EPA-600/2-89-035, 1987a. With permission.)

is the ability of a liquid or gas to dissolve uniformly in another liquid or gas. For example, alcohol and water are completely miscible because of their chemical similarity. Other liquids are only partially miscible or not miscible such as petroleum compounds and water. However, even low water solubilities may still greatly exceed ingestion standards for drinking water.

Vapor pressure can be considered related to the solubility of the compound in air from the liquid phase. The vapor pressure is measured in millimeters of mercury, and in essence is the pressure of the liquid on air, indicating the volatility of the material.

The partition coefficient is a measure of the distribution of a given compound in two phases and is expressed as a concentration ratio, assuming simple dissolution. In reality, the situation could be more complex as a result of molecular changes.

Sorption and desorption, as stated by Moore and Ramamoorthy (1984), means that the more hydrophobic the organic compound is, the more likely it is to be sorbed to the sediment. Absorption occurs when one substance is attracted to and held on the surface of another. Adsorption is that adherence of the atoms, ions, or molecules of a gas or liquid to the surface of another, called the adsorbent (Lewis, 1994). Sorbent characteristics of the geologic matrix include surface area, nature of charge, charge density, presence of hydrophobic areas, presence of organic matter, and strength of sorption. Sorption can be expressed in the equation,

$$Cs = Kp \ Cw \ 1/n$$

where Cs and Cw are the concentrations of the organic compound in solid and water phases, Kp is the partition coefficient for sorption, and 1/n is an exponential factor.

Volatilization is the transport of a compound from the liquid to the vapor phase at a given temperature. This is an important pathway for chemicals with high vapor pressure or low solubilities. A volatile organic compound (VOC) is any hydrocarbon, except methane and ethane, with vapor pressure equal to or greater than 0.1 mm mercury (Lewis, 1993).

Chemical Transformations

Oxidation and reduction reactions (redox) involve the liberation of electrons (oxidation) and reactions that consume electrons (reduction). Many organic compounds can either accept or donate electrons. This is environmentally significant because the oxidized or reduced forms of organic compounds may have totally different physical and chemical properties.

Hydrolysis involves the reaction of hydrogen, hydroxyl radicals, or water molecules interacting with the organic compound depending on the pH and polarity of the reaction site on the molecule.

Halogenation and dehalogenation of organic compounds occur mostly under synthetic conditions or in drastic environments. Moore and Ramamoorthy (1984) state that mild chlorination reactions are possible in natural waters with effluents containing residual chlorine. Dehalogenation may occur under varying reactions of hydrolysis or disproportionation.

Photochemical breakdown processes involve structural changes in a molecule induced by radiation in the near ultraviolet-visible light range. The structure of an organic compound generally determines whether or not a photochemical reaction is possible.

CONTAMINANT PATHWAYS THROUGH THE VADOSE ZONE

Releases from buried tanks, pipelines, building basements, or any subsurface source can promote the migration of contaminants because these locations are below the surface, where direct observation is not possible. This releases material directly in the subsurface vadose zone and can promote vertical movement and decrease transport time if groundwater is shallow. Many recent laws and regulations have addressed the need to monitor potential subsurface release points. A prolonged leak situation may occur for years before discovery, during which a widespread problem has evolved over time.

Contaminant movement through the vadose zone will tend to follow fluid migration according to the movement processes discussed previously. If cracks or fractures (macropores) exist, the contaminant may flow directly into the crack and move downward. This can also happen if the contaminant material is a solid, such as a powered material moved by the wind and settles on a crack-containing surface. This is a very important fluid pathway if the site is covered by clayey soils, which readily desiccate, allowing cracks to form that can be feet deep. Thus, just because the site is covered with a clay (inferring low permeability cover retarding movement) does not preclude a possible rapid fluid or contaminant infiltration movement (see Figure 2).

Movement through the subsurface may be enhanced by biologic structures and voids in soil or sediment. The action of flora and fauna may leave extensive networks of open holes and voids, as well as bioturbating the sediment. Although these voids may only be tenths of an inch in diameter, they can transmit fluid in both horizontal and vertical directions if open to a source of fluid. Additionally, these voids may be open to depths tens of feet or greater in thick alluvial fills, so assuming that voids shrink or close with increasing depth may not always be true. Finally, voids may be infilled with different types of sediment, such as sand in clay, thus increasing hydraulic conductivity in an otherwise lower permeability stratum. Morrison (1989) has reviewed a pesticide transport model to ascertain how the presence of open worm tubes enhances macropore flow (see Figure 3). These tubes allow rapid movement due to "instabilities" where preferential flow may occur as a vertical fingering. The flow in the "fingers" tends to coalesce with depth, so fluid movement is more rapid in vertical directions, with a slow horizontal wetting from the "fingers." Morrison (1989) also points out that while this phenomenon is somewhat texture dependent, it is related to the pore size, number of irregularities (openings), and initial moisture contents. Dragun (1988) states

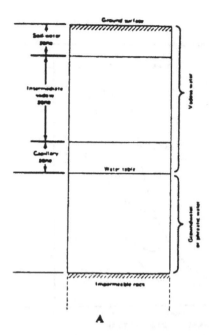

A

Figure 2 Vadose zone. A. Divisions of vadose zone and groundwater zone. B. Hysteresis
— Effect of wetting and drying soil showing increased matrix suction with drying.
(Modified from Freeze and Cherry, 1979 in EPA, 1985. With permission. C.
Unsaturated soil model showing mass transport and retardation mechanisms.
(Modified from Gierke et al., 1985. With permission.)

that bulk hydrocarbons can affect soil hydraulic conductivity by changing inter-
particle spacing between adjacent clay particles. This can cause very large conductivity increases, and allow very rapid penetrtion into supposedly low-permeability soils. The potential to skew vadose model flow assumptions in vadose
zone computer or mathematic models is obvious (see Hatheway, 1994).

Some contaminants may not readily move through the vadose zone unless
some other force of liquid acts on them. In this case, the transport of the contaminants becomes facilitated, or aided by something that enhances movement.
The recharge of water through the soil column may cause leaching, or dissolve
residual contaminant into water that penetrates to the groundwater. Certain pesticides may be formulated as dusts, but when mixed with solvents, are dissolved
and can migrate vertically. This may need to be considered if two or more
contaminants are present, where one is a solvent and the other may be dissolved
by that solvent.

Soil vapor from volatile contaminants can move in the subsurface away from
the source (Schwille, 1988). The movement may occur though natural deposits
or through manmade conduits such as utility trenching, pipelines, into-and-under
structures, and vents to the ground surface or basements. The problems of migrating potentially explosive vapors or formation of unbreathable atmospheres has
been known for some time. However, using soil vapor for subsurface investigation

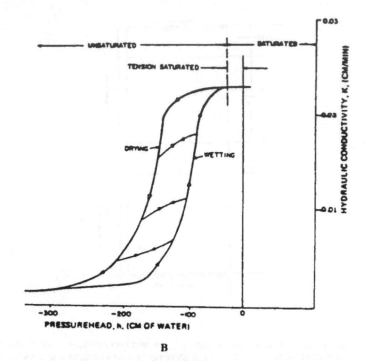

B

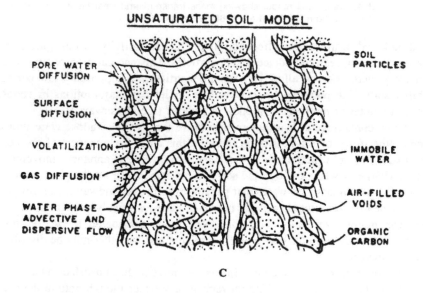

C

Figure 2 (continued)

is a relatively new technique and is widely used as a reconnaissance tool. Basically, soil vapor surveys are done by placing probes 5 to 15 feet into the vadose zone and extracting vapor samples. The vapor is then analyzed in a portable

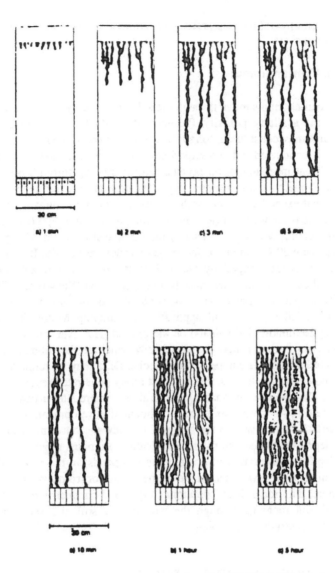

Figure 3 Macropore flow experiments showing wetting front movement in vadose zone. Rapid vertical infiltration occurs when saturation occurs in larger interconnected pores, followed by lateral wetting. (From Morrison, 1989. With permission.)

analyzer (organic vapor or gas chromatography) and a concentration contour map of vapor occurrence is plotted on the site plan. This approach using field mobile laboratories can provide a rapid reconnaissance of a site. The vapor survey will generally "overprint" the groundwater contaminant plume. The isoconcentration contours are used to infer the presence of the contaminant in soil or groundwater, and possible exploratory boring locations are chosen from the vapor survey data. This can save money and time by placing exploratory borings closer to suspected sources to initially define unsaturated and saturated plume extent and monitoring well placement.

Residual Contaminants

When contaminants move through the subsurface, they often leave a residual portion in the available porosity. Presence of residuals are not easily found or removed and may constitute a long-term source of contamination. Conrad et al. (1987) have conducted column experiments examining connection of porosity and pore throat morphology on organic liquids moving through porous media. Their results indicate that residual liquid saturations are influenced by liquid properties that exceed a critical capillary force, and that residual saturations may result from the presence of air (as a nonwetting phase) and larger buoyancy forces and smaller capillary forces. In a sandy medium, average residual organic liquid saturations were 29%, and in the dry range of vadose zone, 9%. If suction is high or contaminants are trapped by water droplets, the contaminant can become immobilized as shown in other glass bead experiments (Schwille, 1988).

Finally, it should be noted that some contaminants may not move very far into the subsurface following the spill, application, or dumping. Motor oil is commonly dumped (by unthinking individuals) on the ground. However, due to its viscosity and organic absorption, it may not necessarily penetrate the subsurface deeply if only a limited quantity is dumped. The pesticide DDT may be found tens of years after use; however, due to its stable nature it rarely penetrates several feet into the subsurface. This includes urban or agricultural use, where later urban development results in testing for pesticides prior to residential development. Trace elements (heavy metals) may move into the soil and groundwater but usually under a condition of depressed pH, so the solute acidification promotes migration.

The depth of penetration of contaminants depends on the type and quantity of contaminant spilled, whether spilled on the surface or from a subsurface source, driving force, duration of leak, contaminant type and chemistry, moisture content of the soil and recharge through the site-specific soil and sediment texture, hydraulic conductivity, and geology.

CONTAMINANT MOVEMENT IN THE AQUIFER

Contaminant migration in the aquifer depends on the properties of the contaminant, aquifer geology, and groundwater velocity. The following discussion will focus primarily upon stratigraphic approaches in either alluvial or sedimentary rock geologic terrains. The three general contaminant migration pathways commonly used in investigation of conceptual aquifer models are: the floaters (immiscible contaminants), mixers (contaminants with uniform dissolution and movement in the aquifer), and sinkers (contaminants that move vertically due to density makeup) (see Figure 4). These general conceptual models are useful in drawing simplified models, but as with all real-world situations, the subsurface

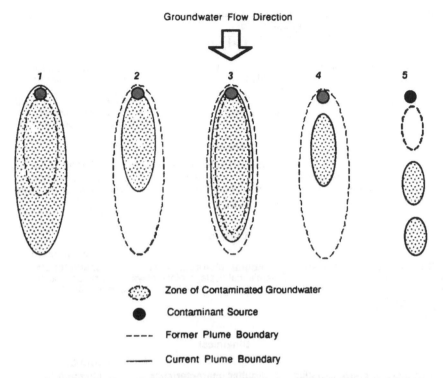

Figure 4 Plume configuration changes and possible factors effecting changes. 1. Increase rate of discharge; sorption activity used up; effects of water table changes. 2. Reduction in rate of discharge; effects of water table changes; more effective soprtion; more effective dilution; slower movement and degradation effects. 3. Discharge rate constant; sorption capacity not fully utilized; dilution effects stable; potentiometric changes not important or not causing change. 4. Discharge ceased; dilution, soprtion, retardation, degradation effects (or combination) shrinking plume. 5. Intermittent or seasonal contaminant source, possible water level fluctuation effects. (Modified from U.S. EPA, 1994. With permission.)

reality is more complex. The individual chemical compound composition, molecular weight, solubility, and viscosity will also influence contaminant movement. For example, hydrocarbon fuels are commonly called floaters, although several common industrial solvents may fit the floater model. Conversely, the sinkers are conceived as industrial chemicals and solvents, whereas all chemicals classified this way may have properties that allow some mixing and sinking (such as industrial oils and saline brines). Generalizations are useful, but may mislead, so the site-specific environments and contaminant chemical characteristics always must be taken into account (see Table 2).

Contaminant Plume Configuration and Movement

Fetter (1988) describes the basic transporting processes for contaminant plume solutes as advection and diffusion. Advection is the process by which

Table 2 Some Information Needed for Predicting Organic Contaminant Movement and Transformation in Groundwater

Hydraulic

Contaminant source	Wells	Hydrogeologic environment
Location	Location	Extent of aquifer and
Amount	Amount	aquitard
Rate of release	Depth	Characteristics of aquifer
	Pump rates	Hydraulic gradient
		Groundwater flow rate

Sorption

Distribution coefficient	Characteristics of the aquifer solid	Contaminant characteristics
Characteristic of concentration	Organic carbon content	Octanol/water partition
	Clay content	coefficient
		Solubility

Chemical

Groundwater characteristics	Aquifer characteristics	Contaminant characteristics
Ionic strength	Potential catalysts: metals, clays	Potential products
pH		Concentration
Temperature		
NO_3^-, SO_4^-, O_2		
Toxicants		

Biological

Groundwater characteristics	Aquifer characteristics	Contaminant characteristics
Ionic strength	Grain size	Potential products
pH	Active bacteria — number	Toxicity
Temperature	Monod rate — constants	Concentration
Nutrients		
Substrate		
O_2, NO_3^-, SO_4^-		
Macro (P,S,N)		
Trace		
Organism		
Concentration		
Distribution		
Type		

From U.S. EPA, 1993c. With permission.

groundwater in motion carries dissolved solutes. Diffusion is the process by which both ionic and molecular species dissolved in water move from areas of higher concentration to areas of lower concentration.

Advection is the rate of flowing water, determined by Darcy's law as

$$Vx = \frac{Ki}{n}$$

where: Vx = average linear velocity
    K = hydraulic conductivity
    n = effective porosity
    i = hydraulic gradient

Fetter (1988) describes the diffusion of a solute through water as determined by Fick's laws and the flux of a solute under steady-state conditions as

$$F = -D \, dC/dx$$

where: F   = mass flux of solute per unit area per unit time
        D   = diffusion coefficient (area/time)
        C   = solute concentration mass/volume)
        dC/dx = concentration gradient (mass/volume/distance)

As the plume moves, processes of dispersion and retardation influence its size and shape. Mechanical dispersion occurs as the contaminated fluid flows through and mixes with non-contaminated background water. The plume shows divergence in the longitudinal and lateral to groundwater flowlines. Dispersion is caused by the differing fluid velocities within the pores and pathways taken by the fluid (Fetter, 1988; see Figure 4). This mechanical dispersion on a microscopic scale is due to a micro-scale deviation from the average groundwater velocity (Anderson, 1993). Anderson reports that several investigators maintain that in order to apply advection–dispersion equations, dispersivity must be defined in terms of the statistical properties of hydraulic conductivity for an aquifer of a given size.

Molecular diffusion occurs as species move from higher to lower concentrations on the microscopic scale (Anderson, 1984). Gillham and Cherry (1982) state that molecular diffusion is important in contaminant transport in fine-grained deposits and may become important in contaminant transport in heterogeneous deposits with low flow velocity, allowing diffusion from higher to less permeable strata. Fetter (1988) states that mechanical and molecular dispersion cannot be separated in groundwater flow regimes. Consequently, a factor termed the coefficient of hydrodynamic dispersion is used to take into account mechanical mixing and diffusion (see Figures 4, 5, and 6). The length and width of the plume will tend to move fastest in the downgradient direction, spreading laterally as a result of the aforementioned factors and aquifer texture (Fetter, 1988; U.S. EPA, 1985). If flow velocity is low, the plume tends to be somewhat less elongated than at higher velocities. If the hydraulic conductivity is low, the plume tends to move slowly and stay relatively compact. Higher hydraulic conductivities may result in more rapid movement and a longer and narrower plume. Overall, Gillham and Cherry (1982) state that contaminant spreading is caused by heterogeneity of geologic materials; diffusion may have more effect in silty-clayey deposits, and regular concentration distributions in sandy deposits may also be more controlled by advection–diffusion.

The plume configuration also depends on the type and chemistry of the contaminant. If it is immiscible, or a light nonaqueous phase liquid (LNAPL) "floater," then a separate phase and dissolved phase will occur near the upper portion of the aquifer (such as gasoline or oil "floating" on the capillary fringe and groundwater surface). For example, separate phase product will migrate to the capillary fringe until the weight of the product overcomes the capillary

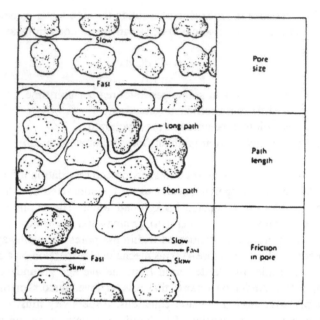

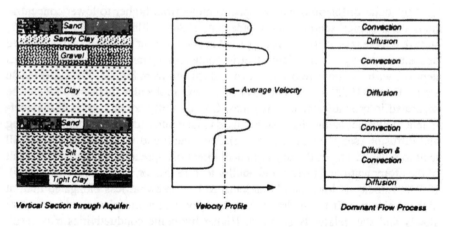

Figure 5   A. Factors causing longitudinal dispersion. (From Fetter, 1988. With permission.)
B. Effect of geologic stratification on velocity and flow processes. (From EPA, 1991. With permission.)

pressure and the product collects and flows on the water surface. When sufficient product collects on the groundwater surface, the surface is displaced downward (Sullivan et al., 1988; see Figures 7 and 8). The separate phase product displaces groundwater entering deeper porosity as groundwater depth fluctuations. A portion of the contaminant will dissolve in groundwater and migrate by advection. These are usually the benzene, toluene, ethylbenzene, and xylene (BTEX) constituents of fuel (gasoline), with the other hydrocarbons contained in fuel (gasoline or diesel) recognized as the total petroleum hydrocarbons. The dissolved and

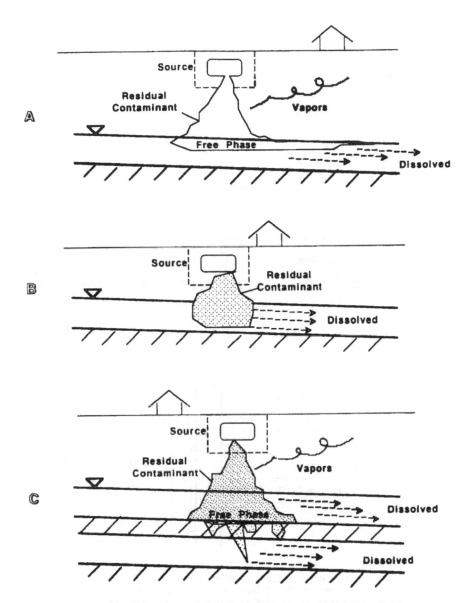

Figure 6 Three ideal contaminant migration models. A. LNAPL or floater (such as petroleum fuels). B. Mixer (such as landfill leachate, acetone). C. Sinker or DNAPL
(solvents, saline brines, PCBs). Free phase liquid may collect given sufficient
quantity of loss in LNAPL or DNAPL. Leakage into underlying aquitards or
aquifers if sufficient DNAPL.

separate phase product tend to be most concentrated in the upper portions of the
aquifer.

Mixers are contaminant chemistries, and solubilities may allow for mixing
(or without some preferential partitioning) through the aquifer. This type of
contaminant plume may become distributed into relatively uniform concentrations

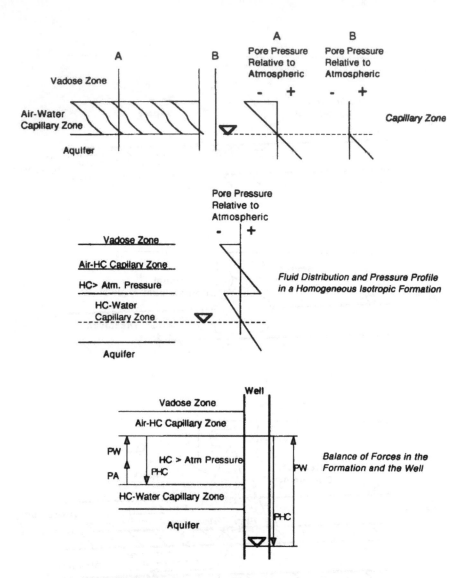

Figure 7   Forces that affect floating product when observed in monitoring wells, HC-hydro-
carbon, Atm-atmospheric, PW-pressure of water, PA-pressure of air, PHC-pres-
sure of hydrocarbon. (Modified from Sullivan et al., 1988. With permission of the
National Groundwater Association.)

once the plume has moved away from the source. A contaminant may have some
preferential partitioning, but if large quantities are released, or it has been in the
aquifer for a long period, concentrations may become somewhat evenly distributed. Municipal landfill leachate is generated from any landfill, past or present.
Leachate may contain nearly any contaminant prior to stricter regulation and even
now, anything could have been disposed in a domestic refuse landfill (hazardous
waste landfills and impoundments may differ). The leachate as a rule tends to be

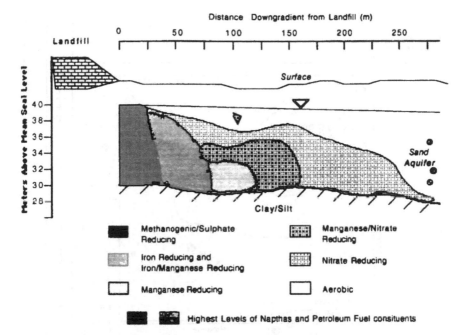

Figure 8   Interpreted distribution of redox zones downgradient of the Grinsted Landfill. Groundwater sampling of the leachate plume from the Grinsted Landfill, Denmark showing the redox variations in the plume. The redox zones are interpreted to be an important chemical attenuation process in the plume. (Modified from Bjerg et al., 1995. With permission from American Chemical Society.)

acidic and may contain both metals and traces of various organic compounds from household and light industrial containers. Studies have observed the leachate migrating from the fill to aquifers, and a relatively uniform concentration distribution usually occurs (Gillham and Cherry, 1982). Recent work by Bjerg et al. (1995) shows that different areas of redox zones can exist in leachate and affect contaminant movement and attenuation processes (see Figures 9 and 10).

A "sinker," or dense nonaqueous phase liquid (DNAPL), contaminant plume may display a concentration gradient through the aquifer, becoming more concentrated near the aquifer base. If a sufficient quantity of sinker enters the aquifer with sufficient driving head, a separate phase may collect at the aquifer base. Sinker contaminants may include the widely used industrial solvents (TCE, PCE, TCA), but can be other materials, such as saline brines and polychlorinated biphenyl (PCB) oils. Industrial solvents that were widely used in manufacturing and at military sites are denser than water (volatile chlorinated hydrocarbons, or CHC) Model column experiments by Schwille (1988) indicate that CHCs will sink given enough CHC fluid pressure in unsaturated and saturated media. CHC penetration occurs after it has developed sufficient head to drive out water (in saturated media), and it is possible that CHCs will not effectively penetrate moist and heterogeneous soils due to oversimplified assumptions of CHC density and viscosity properties (Schwille, 1988). These DNAPLs may become immobilized

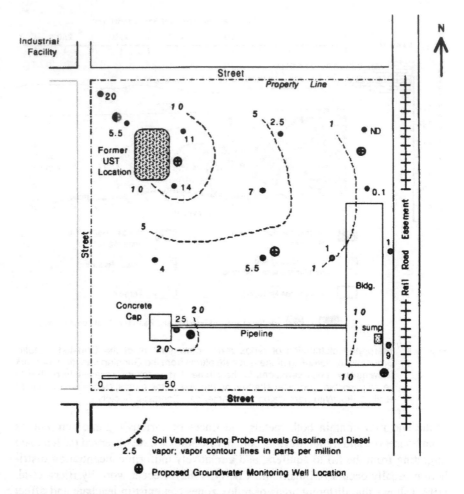

Figure 9   Initial plume location. The site was first field screened with soil vapor mapping.
The client reported that the former UST location had gasoline and diesel fuels,
which are his primary concern. The soil contour vapor data shows possible
sources at the concrete cap and former UST but elevated levels occur in an
apparent upgradient location. Vapor data near the railroad could be incidental
contamination from the railroad or from on-site sources. The groundwater monitoring wells are arrayed to map possible plume on-site, with one well upgradient
of the former UST to look for an off-site, one possibly off-site, and the sump and
railroad easement are suspect sources. Monitoring well locations are chosen
appropriately.

in saturated porosity and be a long-term source of dissolved contaminant. Recent
column experiments by Abdul et al. (1990) were done to evaluate the flow of
organic solvents through kaolin and bentonite. Their results indicate that aqueous
solvents did not change the physical appearance of clays, and solutions moved
through the clays at a constant hydraulic conductivity. However, hydrophobic
solvents caused clays to shrink, forming networks of cracks, allowing fracture
flow to occur through the clay (see discussion in Brusseau, 1993).

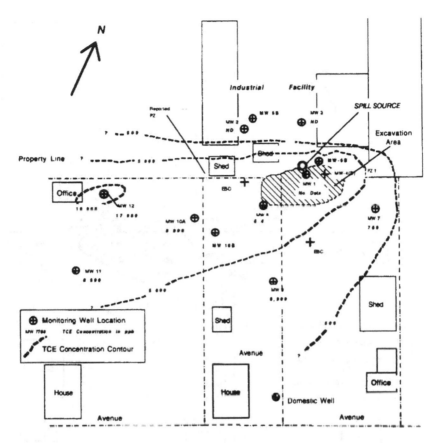

Figure 10 TCE contamination mapped in the "A" aquifer by exploratory borings and monitoring wells.

Contaminants may partition or separate once dissolved into groundwater. For example, gasoline (a fuel that may contain tens of individual compounds) may partition into benzene, toluene, xylene, and other hydrocarbons; this type of partitioning is often observed at fuel spill sites (see Figures 9, 10). CHCs such as trichloroethylene may similarly partition into dichloroethylene breakdown products that migrate at different velocities. Complex mixtures of groundwater contaminants may occur if mixed contaminants are present (Brusseau, 1993). For example, an enhanced or facilitated transport may occur from relative solubilities of low polarity organic solutes in organic liquids and water, or decrease when an immiscible liquid is present as a fixed residual phase. For example, Brusseau (1990, in 1993) observed retardation of naphthalene in a column of aquifer materials with a residual phase of tetrachloroethene.

Fetter (1988) states that there are two broad classes of solutes, conservative and reactive. Conservative solutes do not react with the soil or groundwater, nor do they undergo decay (for example, chloride). Reactive substances can undergo chemical, biological, or radioactive change (degradation), which tends to decrease the concentration of the solute. If a contaminant undergoes degradation, a single

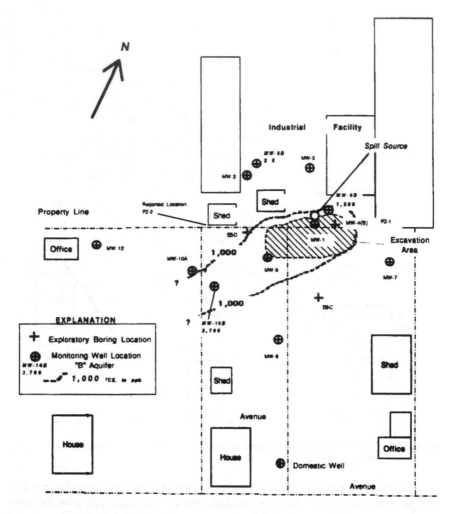

Figure 11   TCE in the "B" aquifer. Compare to the "A" aquifer map and note that concentrations contours hook around the spill/excavation location and the movement
is to the southwest across property lines.

reaction or a series of degradation reactions may occur. For example, experiments
by Barker et al. (1987) show that benzene, toluene, and xylene will migrate at
different velocities and will undergo biotransformation and degrade given sufficient supplies of oxygen in the aquifer. Other degradation processes may include
temperature, microbial processes, oxidation and reduction, instability of the compound in question, and reaction to form a new compound (such as chloroform
or trihalomethanes). Further, the breakdown may proceed to methylene chloride
which is more "toxic" than the original trichloroethylene contaminant (see Figures
11 and 12).

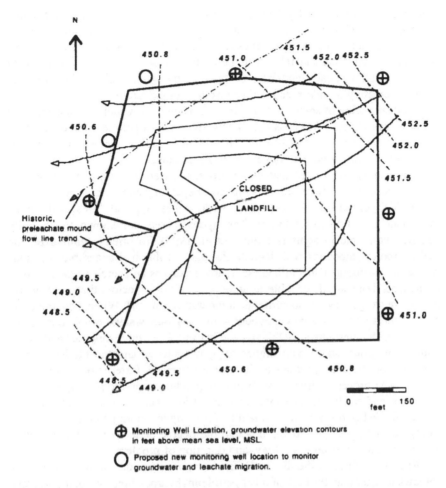

Figure 12   Monitoring at a closed landfill. A leachate mound appears to have developed in the northwestern portion of the landfill. Two locations are selected based on the regional flow gradient, which is consistantly to the southwest and the newly formed leachate mound.

Effects of Aquifer Stratigraphy on Contaminant Movement

The occurrence of sand and clay (higher or lower conductivity units) and their bedding and contact relations will profoundly affect movement in the aquifer. For example, thin clay beds may split, retard, or deflect flow, causing the plume to spread horizontally and vertically. The pore size change of fine sand interbedded with sandy gravel, or coarse-grained sand may create "capillary gaps" where the downward migration of contaminant in the vadose zone may slow and spread laterally. If the contaminant fluid tension is too high and/or driving head is too low, the fluid may not enter the larger pores of the underlying unit. Similarly, if the porosity is clogged by fine-grained matrix, the plume may move slowly, and move at lower flow rates than calculations may imply. Sand lenses may allow more rapid movement. Clay between coarser grains, or as strata, may become absorbed to the clay. This can retard movement and immobilize the contaminant, but can create a residual contaminant source, releasing it slowly into groundwater. As groundwater rises and falls in the aquifer, a floater petroleum contaminant can be spread vertically and later be trapped by water refilling the available porosity, creating a widespread source throughout the aquifer.

Biodegradation of some contaminants has been more widely recognized over the past several years. This has been known to occur in petroleum hydrocarbons for some time and is used to effect site remediation, research is ongoing for other hydrocarbons, and a brief summary of it is presented.

It is widely recognized that hydrocarbon fuels are mixtures of compounds that may reach up to 200 in some fuel blends, depending on refinery and company. The BTEX compounds are of interest because these typically present the most immediate threat to groundwater. The hydrocarbons are known to naturally degrade with time in aquifers (Norris et al. 1993). To briefly summarize, the hydrocarbon compounds will degrade through aerobic and anaerobic reactions where the microorganisms utilize the hydrocarbons as an energy source. Almost all hydrocarbons are degradable in aerobic conditions. Factors that limit degradation are oxygen, available nutrients, temperature, and pH. When a contaminant plume enters and moves downgradient with the groundwater, aerobic conditions will commence until the available oxygen is used. The plume migrates and mixes with more groundwater with higher oxygen contents, and the biodegradation continues on the outer portions of the plume somewhat faster than on the plume interior. The process continues with time depending upon the geologic environment type and can be enhanced if adequate nutrients and oxygen can be delivered to the affected portion of the formation. This implies more potential success in sandy rather than clayey strata, although the degree of success and degradation are site specific to source, geology, and duration of problem.

As a general rule of thumb, the BTEX components of the plume may migrate somewhat faster than the bulk of other petroleum hydrocarbons. Also, the benzene and ethylbenzene tend to degrade more quickly than toluene and xylene (see Barker et al., 1987). BTEX will degrade over time and concentrations will asymptote in the range of parts per billion to tens of parts per billion. The long tailoff of concentrations is partially due to the degradation reactions and residual drainage of contaminants into and migration through the aquifer.

Misinterpretation of Contaminant Location by Oversimplifying Stratigraphy

Inferring a simple "layer cake" stratigraphy and utilizing overly simplistic homogeneous conditions can be quite misleading. Conceptual flow paths must be based upon the actual site geology and hydrogeologic constraints. Internal stratification layers' thickness and lateral extent must be located because they can deflect movement laterally or enhance vertical movement. This relates directly to the accuracy and completeness of lithology logging collected in subsurface investigations, especially the presence of mixed textures, stratification and fractures or routes of secondary permeability. Kueper et al. (1993) have shown in sand box experiments that internal stratification may affect the movement of tetrachloroethlyene and its ultimate dispersal in sediment, including moving laterally on bedding planes. Again, the existing background chemistry and introduced contaminants must be factored into the conceptual geologic and groundwater flow model so that both field investigations and data analysis are meaningful.

CONTAMINANT MOVEMENT IN FRACTURED ROCK

Contaminant movement investigations can be very complicated in geologic terrains containing fractured rock and sediments. The fluid movement through the fractures makes aquifer analysis and modeling contaminant movement very difficult. Schmelling and Ross (1989) have prepared a review of contaminant movement in fractured media, from which the following is summarized. Fluid movement in most fractured rock systems is through fractures, joints, cracks, and shears that may occur in sets or zones. Fractures may be open or infilled with mineralization. Shears or faulting movements may produce gouge or slickensides, which may form "skins" that can retard fluid movement. Groundwater movement will depend on the fracture density, interconnections, orientation, aperture width, and nature of rock matrix. Igneous and metamorphic rock usually have low primary porosity and permeability, so fractures form the primary fluid pathway. Sedimentary rocks may have higher primary porosity and permeability, and fracture permeability is secondary; however, depending on cementation and induration, fractures may form the primary permeability.

Flow through fractures may be through a fracture system or through a few fractures that predominate flow (Neretnieks, 1993). Water mixing may occur at fracture intersections and this may have a tendency to average out concentration. Schmelling and Ross (1989) state that rates of contaminant migration into or out of rock will depend on matrix permeability, presence of low-permeability fracture "skins," and matrix diffusion coefficient of the contaminant. Parker et al. (1994) have shown that DNAPL may move in fractures in an immiscible phase and diffuse into the aquifer matrix porosities. Thus, DNAPL or LNAPL may become "cut off" in dead-end fractures and desorb low concentrations to groundwater for long periods of time, making cleanup a very difficult and long-term process.

CONTAMINANT MOVEMENT BETWEEN AQUIFERS

Aquitards that separate individual aquifers are impermeable in the conceptual sense only, and transmit fluid, albeit very slowly. Toth (1984) shows that while hydraulic conductivities of clay or any low-permeability unit are low, flow will occur through these units. Additionally, the geochemical nature of the water will change due to the thermal and chemical interaction during flow. Groundwater geochemistry will show changing composition of water moving naturally when modeled in Toth's Unit Basin (Toth, 1984). Hence, natural chemical changes occur as water migrates through the aquifer and aquitard system. Natural leakage through aquicludes and aquitards by leakage at strata contacts or through fractures allows groundwater to move and can significantly change the aquifer water geochemical character. When contaminants enter the system, they may move as a natural chemical system would and in sufficient quantity and driving force, move through aquitards. However, movement can be enhanced since they could be introduced at any depth and in any geologic unit. The most recognizable case is through a well that interconnects several aquifers, creating an opening through which groundwater may be transmitted vertically (see Figure 1). Contaminant movement can be accelerated by pumping wells that draw the contaminant toward the well while mixing contaminated with uncontaminated water.

Wells may have been used for illegal disposal — dumping unwanted material down the well casing. Since regional well records may not be accurate or complete, exact well locations are unclear or forgotten, and the driller's log may be incomplete — lacking lithologic data and aquifer contacts definition. Construction details are often incomplete or missing. Often wells are abandoned with pumps and lines in the casing, which rust and may collapse, precluding future sampling. Finally, agricultural wells may have screened several aquifers for maximum yield, interconnecting several water-bearing zones. These farms and agricultural wells are sold and developed through urbanization of the area. If the land is rapidly developed, wells may be buried and forgotten, leaving many points of aquifer interconnection in industrial or residential areas (including geotechnical engineering borings used for foundation studies). Considerable analysis and effort may be required to relocate those wells. Older well casings can corrode, and the boreholes may collapse but still allow water movement between different strata. Once relocated, the well must be properly cleaned and reconstructed, or abandoned and grouted shut.

LOCATING THE CONTAMINANT PLUME — THE INITIAL APPROACH

In order to attempt to locate the plume, the hydrogeologist will have to make some preliminary assumptions. Assuming the contaminant is present in the aquifer, the next assumption is that the groundwater chemistry is affected and moving at some average linear velocity. The hydrogeologist may review nearby previous investigations reports, soil vapor surveys, surmised contaminant properties, and any other available data. At times, surface geophysical techniques may aid in initial location of suspected contaminants (such as electromagnetics or surface resistivity). Ultimately, a subsurface investigation will be conducted to provide the geologic data, sample media for chemical analysis at specific depths, and install the permanent monitoring points. The money and time that will be committed to the subsurface study may vary depending on the level of effort, and typically the investigation is phased so time and money are more effectively spent; it is unrealistic to expect complete definitive answers from only one study. Usually, the basic geology and aquifer system and contaminant chemistry information must be collected first for an initial site appraisal.

Exploratory soil borings will be drilled and monitoring wells installed to collect the basic data. Some boreholes may be advanced to collect lithologic and chemical sampling information only, which are used to locate the monitoring wells. However, if groundwater gradient and flow direction are to be known, at least three monitoring wells are required. If this is the initial site investigation, then one well is placed in the surmised upgradient direction at the source and two are placed in the surmised downgradient direction. These wells will allow for repeated measurements for gradient calculation and sampling with additional wells are installed to detail the hydrogeologic and chemical data requirements. Groundwater contamination investigations can be long affairs and it is possible that numerous wells will be installed to adequately define and monitor the plume (cross-gradient and downgradient). Complications may arise if wells cannot be positioned in the most desirable locations due to access or legal problems. This may mean long-distance extrapolation of subsurface data if drilling access is not available. Finally, the geology may or may not be easy to resolve or interpret, which requires additional subsurface work. The monitoring wells will provide the long-term sampling for the periodic monitoring that will be required to evaluate the completeness of site characterization and cleanup effectiveness.

Ultimately the number of wells and level of effort required to define both unsaturated zone and groundwater contamination largely depends on what is revealed by the initial field studies. A complex project can require several field phases, but the goal of all data collected from each step should be for contaminant delineation and cleanup. This can be especially true of a widespread contaminant problem, or a DNAPL problem in multiple aquifers. Some site screening techniques may be used as soil vapor mapping, or drilling some boreholes for reconnaissance sampling, but this does not replace the need for permanent monitoring wells. Although the effort may take the appearance of a "research project" to the client, it is in fact an investigation to solve the problem, no more and no less. The amount of scientific effort and analysis must be sufficient to get the required answers for site characterization. It is up to the consultant to help the client visualize the investigation goals and how they benefit his cleanup activity.

OVERVIEW OF THE REPORT AND DATA ANALYSIS

Once the field work is done, the information must be organized and analyzed to determine whether or not additional work is needed to complete the investigation report and begin remediation. Some type of formal reporting document will be needed at this point. Some regulating agencies may want formal reports and others may require only memos updating them on site activity. If a state reimbursement fund is involved, then reporting and documentation of budget spent are sent to the agencies to document the amount of reimbursement.

Each consultant formal letter or report becomes a legal document, which the responsible professional must defend technically and legally and to the highest ethical standards (see Association of Engineering Geologists, 1981). There are times when the investigation is only preliminary or a reconnaissance detail level. If this is the case, that report should clearly state the limited detail of scope. This may become an issue with implications regarding completeness or omission, unless the report actually states the limited, or comprehensive scope of work. Thus, preliminary results may indicate a cleanup approach that may or may not be suitable for that site condition.

Usually local, state, and federal regulatory guidance documents will set content requirement for a written report. It has been the author's experience that the following sections are typically included in most reports in order to present the subsurface information in a format for delineating site geology and contaminant extent. These parts would comprise the minimum content, with other sections added as needed to explain or address all the investigation issues. Hence, the following sections are usually included:

Site plan or map — shows monitoring well locations, site surface geology and geography, site improvements (utilities, equipment, tanks, etc.) and any other required information. Several maps may be needed to represent the information. The maps should show the necessary information and cite sources of previously obtained information, especially if maps, tables, or cross sections are adapted from a published or unpublished report. Maps derived from preexisting sources should contain the reference on the map proper.

Cross sections —— shows the site stratigraphy, aquifers, boring locations, well details, and other relevant geologic and hydrogeologic information. The geologic data on the section may be generalized to a point for portrayal purposes, but should not compromise accuracy. Often, the site stratigraphy shown is the "hydrostratigraphy," where the strata of interest are portrayed in relation to contaminant movement in aquifers or aquitards. The surface and well casing elevations should always be set to depth relative to mean sea level. Chemical data is often included, especially from vadose zone chemical sampling. All vertical and horizontal changes in stratigraphy and contaminant changes should be shown using as many sections as may be required.

Groundwater elevation contour maps — display the groundwater contours and flow lines for the day the data were collected (typically the date on which the wells were measured and sampled). The groundwater elevation contours should always be referenced to mean sea level.

Groundwater contamination contour maps — shows the distribution of the contaminants in water in relation to the site boundaries and contaminant source. Concentrations are often shown in logarithmic concentration contours since the values are commonly analyzed and reported in parts per million or billion. The concentration contours can also represent data from cross sections in the vadose zone and aquifers where appropriate.

Data tables and appendices — a large quantity of chemical and other data is generated by the investigation. A table format makes the data readable and concisely presented. Presentation by well and date of sampling is usual, and can be referenced to the narrative, figures, and graphs. Appendices usually contain the exploratory boring logs, well construction details, chemical analytical reports, permits, correspondence, and anything else that supports the body of the report.

Report narrative — will include all methods, procedures and results of the study, along with the backup documents. The basic sections will include the geology, hydrogeology, extent of pollutants, and interpreted condition of the affected area. This may or may not have input from other consultants and clients and even lawyers; however, it is up to the responsible consultant to adhere to his interpretations and report completeness. The grammar used, and the words chosen, should convey what the site study has revealed clearly and without ambiguity. Accepted geologic and hydrogeologic terms should be used to describe the subsurface where needed to adequately describe the geologic condition or item of interest. Technical words should be used where needed and explained to the layman level, but the report should not be "jargonistic." Where the opinion of the responsible professional is given it should be stated as such (in other words, the data observed is then interpreted). The definitions of words may cause problems in the future, and so the choice of words may become a point of debate (e.g., "observe" rather than "examine," or "suggest" rather than "indicate"). This may seem to hedge about the issue (the so-called weasel words), but they are unfortunately a fact of life when considering legal documents. Consider that you are working within another framework of words and laws in regulations, some of which may be ambiguous yet demand technical explanations to address those regulations. A large portion of the audience may be less familiar than the consultant. The wording should not confuse or compromise the basic scientific integrity of the work or its conclusions. The report may also have recommendations sections, but the data and work must support both those conclusions and recommendations in all cases. Finally, all citations to other work, especially other consultants' work, and appropriate legal regulations should be included both in the text and on separate citations or reference lists.

The consultant author should ask himself if all the information makes sense. Do the exploratory boring logs substantiate the correlation of soil/alluvium/rock units? Were a sufficient number of soil and groundwater samples analyzed? Do groundwater maps and elevation contours agree with local and regional flow concepts; if not, why? Are chemical results in agreement with observed trends or expected values given the hydrogeology and contaminant type? Do the monitoring wells and chemical data accurately locate the plume? Where are the data gaps? Which issues are resolved and which are not? This review and questioning should take place throughout the investigation and report drafting. Technical issues should be anticipated and the consultant should answer questions regarding plume extent and regulation before the report is issued.

It has been the author's experience that the maps and cross sections are typically the most reviewed portion of the report, whereas sometimes people do not carefully read the accompanying text. Usually the presentation format is similar to the geologic and civil engineering presentations adapted for use in contaminant hydrogeology. The maps then should be carefully prepared and the data must be accurate inasmuch as maps and sections are sometimes reviewed without a careful text perusal. Although the text is the most complete description of what was done, the maps and sections are shortcuts used by all interested

parties, consultants, regulators, and clients. There is always the possibility of
these documents being used out of the text context.

Each investigation is different given the type of problem, and geology will
always differ from site to site. Even though similarities may exist in sites and
geology, one can never apply the same solution to different sites and assume one
can obtain equally effective remediation. The information derived in the investigation and testing is valid for that unique site and must support its site-specific
conclusions.

Comments on Using Previous Investigation Reports

It is very possible that a consulting hydrogeologist will finish a study started
by another, or be asked to comment on another consultant's work. When con-
ducting or working an investigation on which another consultant has worked, you
should be careful to be ensure that the information is accurate and agrees with
your work. Often you must refer to reports done by other geologists or consultants
and agencies and must evaluate the validity of the work as it relates to your site.
If you accept work that is inferior or flawed, you will compromise the accuracy
of your work. Several problems may arise in using other consultants' work. The
following examples may be encountered but are by no means a complete list and
the hydrogeologist must always guard against accepting dubious information.

Problems can arise from an incomplete definition of the extent of vadose
contamination. If this is incomplete, a residual contaminant may remain that will
leach additional contaminants into the groundwater body. This will cause future
problems if the contamination remains in the ground for some time, or additional
costs when remediation is attempted. The regulating agency may ask for more
definitive work and remediation. Obvious costs and legal problems will arise due
to incompleteness of a report.

Another series of problems arise from cross-contaminated aquifers as the
result of previous well installations. When wells (or exploratory borings) connect
several water-bearing horizons, the potential for interconnection exists. The strata
may not be saturated when drilling, yet when seasonal precipitation recharges
the dry strata, possible cross connection could occur by remobilizing contaminants. Contaminated water may move vertically in the interconnection, spreading
the problem. This may occur if the lithologic logging was incomplete, sloppy, or
not done, as is the case in many agricultural wells. Lithology logging styles and
completeness vary on the exploratory boring, and useful hydrogeologic information may not be included in an inferior boring log. Contact identification of
aquifers and aquitards may be vague or only estimated during drilling and may
be inaccurate. The well design detail should be reviewed to check to see if the
designed screened interval is contiguous with the supposed aquifer, and sealed
into the aquiclude. If the construction of the well is sloppy, seals and sand packs
may not be located at the intended design interval.

Another problem is incomplete definition of the so-called "zero line," or that
the furthest observed extent of the groundwater plume was not detected. If the
extent of the plume is not defined by chemical data sampled from boreholes and

wells, regulating agencies could require more work and add cost to a project. More importantly, an incomplete definition of the plume may skew the remediation effort, and not completely capture the contaminants. Thus, portions of the plume may continue to migrate, creating complex cleanup and liability problems.

A final problem involves the quality of all chemical analytical data, especially that from previous investigations. The data may be suspect if sampling and quality assurance and control procedures were not properly accomplished. Sampling and sample-handling procedures should be reviewed in light of the reported analytical values. The recorded analytical data should be reviewed in light of the laboratory procedures for accuracy and reproducibility. Inferior chemical data may lead to erroneous interpretations and conclusions, which may compromise existing effort and direction of the next phase of the study. Good chemical data is crucial for placement of exploratory borings and monitoring wells used to delineate the plume.

EXAMPLE APPROACH TO A SITE INVESTIGATION

You are retained to investigate an underground storage tank (UST) site. The site has an industrial history with one known UST, one suspected UST, one sump and is located in an area of long-term industrial use (see Figure 9). The client has limited funds for site assessment, yet wished to make a "good faith" effort to define the problem with the minimum expenditure.

A site reconnaissance is in order to attempt to locate the monitoring wells at the UST and try to make a best guess of groundwater flow direction, if a problem appears to exist. You elect to use a soil vapor mapping approach as an initial approach since the UST liquids are volatile and reported to be fuels. The area of the concrete cap looks suspiciously like a former UST so some effort needs to be expended there. A railroad track behind the site has been used for loading and unloading for years, so some effort is needed there also.

The site is mapped as shown in Figure 9. The results of the soil vapor survey are plotted and show the distribution presented. Soil vapor data reveal an apparent plume moving to the east. There was only enough time and budget for two probes of the concrete cap, so one was placed near and one far. Because a plume was mapped at the known UST, the second could be added to groundwater monitoring for the first. It appears from the soil vapor data that a larger problem may exist at the concrete cap, and a plume may be moving from across the street to the northwest. Groundwater monitoring wells are proposed at the locations shown to provide as much coverage of the site as possible, with upgradient and downgradient monitoring.

Mapping a TCE Plume in Two Aquifers

An industrial site had a release of TCE from a tank where the TCE was used in cleaning processes. A large volume of TCE was released and spread and percolated into the ground, even though a clay soil capped the region. The problem appeared severe enough that the industrial client elected to attempt to excavate the solvent before there was time to create a groundwater problem in the shallow aquifer. The excavation continued until the lateral extent of the apparent vadose contamination was removed and groundwater was encountered. The soil was treated and removed, and clean porous fill was used to backfill the excavation. Groundwater wells were initially installed to map the immediate area, and revealed a clayey aquifer, which was assumed to allow only slow groundwater movement. At this point, the client (lessee) and second party (property owner) began to argue over costs and responsibility. Action on the site languished for a year.

The second consultant retained was to map the site groundwater plume in the upper aquifer and ascertain the flow direction and presence of TCE in the second underlying aquifer. A budget of $250,000 was thought to be sufficient to map the entire plume, assuming that TCE would not move quickly in a clayey aquifer and inasmuch as the immediate spill source had been removed. The monitoring well installation was delayed for several months because access to several properties was needed and the negotiations required legal agreements to be approved by all the parties.

The site was mapped with the monitoring wells as shown in Figures 10 and 11. The results of the groundwater plume mapping revealed some unpleasant aspects of the site. First, the TCE dissolved in the upper "A" aquifer showed very high levels and the groundwater plume had moved quite far away from the spill source. The plume was dispersing downgradient, and the geology revealed the the aquifer became sandier to the west-southwest. A "slug" of elevated TCE levels was moving to the west, and seemed to confirm that a one-time release had occurred, and movement may be along a preferred pathway in the aquifer. A house to the south of the spill was using an on-site well for domestic water and the plume appeared to be moving in that direction.

Three monitoring wells placed in the next underlying "B" aquifer revealed more bad news. A large plume was moving away to the west, at concentrations of 1000 ppm, centered on the spill source. This implied that sufficient quantity of TCE had sunk into the underlying aquifer, and a dissolved plume was migrating away. The aquitard was as thick as 10 feet, but the TCE had broken through. Although aquitard soil samples indicated that the TCE penetration was minimal, those samples were not taken directly in the backfilled area since the wells were positioned due to utility constraints. The plume in the "B" aquifer was not completely mapped either, and appeared to be moving away in a confined plume in the sandy and gravelly aquifer.

In this case, your consulting groundwater mapping work is fine as far as it goes, but the realities of this problem are stark. First, it appears that a large quantity indeed of TCE escaped, perhaps larger than the client initially estimated. Second, a vadose cleanup was acceptable as far as it went, but the groundwater should have been investigated immediately to ascertain contaminant penetration. Next, bickering over money and site property access delayed the work, and although unavoidable, the groundwater plume continued to migrate away from the spill source. Finally, and worst of all, the plumes in the two aquifers were only partially mapped since the supposed aquifer texture of clay in the upper aquifer was assumed to allow only relatively slow contaminant transport. Now the groundwater plumes appear to be only half mapped and the entire anticipated exploration budget has been expended. Because a domestic well is threatened, more work must be done, and soon, to complete plume mapping. Even though delays may hold up work, the plume continues to advance and the parties' liability will grow unless drastic action is taken, a tough choice for the client and property owner.

EXAMPLE OF MODIFYING THE MONITORING SYSTEM AT A CLOSED LANDFILL

A closed municipal landfill has been closed for several years, according to the prevailing regulations, with proper capping and security. The city sent this project out for bid to provide monitoring services and consultation for the landfill be be sure it is in regulatory compliance. Your firm has won the job and your first review of the groundwater conditions is that of the groundwater flow (see Figure 13). Several years of data have established a groundwater flow gradient to the southwest with the existing array. However, a leachate mound has developed in the northwestern portion of the landfill. Consequently, groundwater flow lines suggest that the existing monitoring system may not provide the required monitoring coverage. Two additional monitoring wells need to be installed to monitor possible flow paths. A chemical analytical program would be the same as previously, although indicator parameters may increase if leachate and depressed pH have collected and started to migrate; it is anticipated that northwestern wells would observe this in the northern portion of the fill before westerly wells. Since the municipality is cash short, minimal monitoring that addresses the issue should be a workable solution.

SUMMARY

An initial plume definition approach should give the hydrogeologist the basic information required to form valid conclusions. The geology, aquifer, and aquitard should be identified as well as groundwater occurrence and flow direction. The chemistry of soil samples should yield the minimum extent of vadose vertical and horizontal contamination. Groundwater data should show the contaminants distribution on site with some indication as to whether the contaminants have moved off site. The upgradient well should answer whether solvents are moving on-site from an off site source. The report should accurately and concisely report all information, using the appropriate technical standards and formats required by the regulating agency.

Unanswered questions will remain following the investigation, since only rarely are all answers found from a single study. Another phase of study may be required so the hydrogeologist feels the coverage is adequate. The data generated will be tested by the degree and extent of extrapolation from data point to point. The hydrogeologist must decide just how far the data can stretch given the budget without compromising quality. The hydrogeologist must give the client his best opinions, most scientifically accurate report, and interpretation of the data, subject to budget availability and legal requirements. This sometimes means delivering "bad news" to the client that implies spending more money. The client may not wish to spend additional money. Legal implications of whether or not to conduct more investigation are ultimately the client's responsibility. Although the hydrogeologist may do his best, the data may not be complete and the client may elect not to budget additional work.

CHAPTER 3

Subsurface Exploration, Sampling, and Mapping

INTRODUCTION

Prior to starting a subsurface field study program for a groundwater contamination project, several tasks should be addressed. These involve getting the proper people and equipment to the site, ascertaining the source of the contaminant, and meeting regulatory reporting requirements. The responsible party is usually contacted by an agency, or will do the work in response to a spill, hazardous materials storage leak ordinance, or some agency order.

Initial work scope decisions take into account the project goals and what type of investigation activity is undertaken, as well as budget and time constraints. Usually the consultant is working under some kind of client contractual arrangement in which scope of work is defined under an itemized budget. Thus, some type of bidding document has been prepared for the work plan. Ultimately, the work plan is reviewed by the regulating agency, depending on location and how the regulations have evolved, or if state financial reimbursement programs are funding the work.

The technical work plan could be voluminous for federal RCRA/CERCLA work, or could be a three-page letter for an underground storage tank (UST) study. The site work will usually have some kind of requirement of sampling, analysis, and interpretive narrative to address the site problem. However, the work plan must be properly executed in the field. Logistical considerations peripheral to the technical conceptualization are as important as the field data collection since the staff and contractor equipment must access the site. In other words, the hydrogeologist must know how to schedule, plan field operations, and select contractors for the project. This chapter reviews some of these considerations and how they relate to performing the technical and scientific portions of the investigation.

SUBSURFACE EXPLORATION PROGRAM APPROACH

Geologic Data Collection Goals

The subsurface program is used by the consultant to collect the information the regulatory agencies require. These regulatory agencies (federal, state, or local) will require information on the vertical and horizontal extent of "soil" and groundwater contamination at the site. The program should be designed to address the following:

1. establish the overall site geologic setting;
2. allow definition of the site hydrogeology, aquifers, aquitards, groundwater occurrence, and flow directions;
3. sampling "soil" and groundwater at discrete intervals or locations in both unsaturated and saturated zones for both contaminant and geologic data;
4. the physical and chemical testing parameters for the project.

Thus, the hydrogeologist needs to outline the coverage of the aforementioned needs for site delineation prior to going into the field.

This approach should yield a clear understanding of the site conditions, especially the relationship of the geology and hydrogeology to the contaminant type and extent. It is very important to remember that this implies a completeness of information collected for the available budget. The available money may limit the number of sampling points, but it should never compromise the information quality and accuracy.

The consultant should always use all possible information sources at his disposal to gain local site data for incorporation into the work plan. The primary sources should be the site owner, who has the site history of use, industrial practice, waste disposal history, materials manifests, and pertinent regulatory correspondence. Background geologic data such as USGS, state geologic survey, soil survey, and water resource reports and local unpublished groundwater studies should be reviewed where available. City and county planning offices may provide reports on property development phases. These offices may also have geotechnical reports containing exploratory boring logs that detail previous observations of strata, groundwater occurrence, and occasionally reports of unusual chemical problems. Regulatory file review is almost a mandatory step today because numerous parties investigate these problems. Unfortunately, regulatory files are not necessarily complete, or reports are not filed in a timely manner, so even though reports are not in the file, it does not preclude work having been done on site.

Field Logistical and Drilling Contractor Selection Considerations

The consultant is responsible for getting the people and equipment to the site for the field work. There are always logistical problems associated with field work, whether the site is downtown or on a remote hillside. Numerous considerations and problems may arise including site access, borehole and well permitting, site utility clearance, and the execution of the work within the time and budget limits. These can have a direct bearing on the success of the project.

A project requires an approved budget, so the consultant will spend the client's money. Although an emergency response might be done "without cost considerations," the usual case is to operate within budget constraints. Consequently, each portion of the project should be cost estimated. Figure 1 shows possible cost items that could be involved in an investigation. The consulting company is a business, and a company profit will be figured into that total budget. The client will review the budget and scope of work presented in the consultant's work proposal. The client must deal with the pollution problem, yet wants the most for the money. Thus, the consultant must provide the maximum information for each dollar spent. This implies that the budget may constrain the information coverage available for the data interpretation and problem definition.

Site access and permits needed for subsurface work are usually combined tasks since they precede the actual field work. Often, well installation permits are needed for monitoring wells just as they are for water supply wells, and will provide documentation of well location and construction. These permits may be required by the local oversight agency (LOA) and require some kind of approval prior to starting work. In some states, the permit review is a cost item and funding vehicle of the LOA review, and needs to be budgeted. A time delay may occur during the review process and a time schedule of anticipated work task completion is included. Other access permits include encroachment, subsurface work in public easements, right-of-ways, and so on, which can require letters of authorization, fees, and scheduling field inspectors for the work. These regulatory fees may be billed directly to the client.

Site clearance involves marking overhead and subsurface utilities and the site safety plan. Buried and overhead utility lines must be marked and avoided to prevent injury and costly repairs for the field crew and especially the drillers. The site safety plan must be prepared and reviewed for appropriateness and changed conditions if encountered. Safety plan preparation is beyond this book's scope, and the reader is referred to the appropriate state and federal regulations for the safety plan; however, a brief review follows. The site safety plan must meet the basic state and federal requirements for worker protection (drillers, geologists, engineers, samplers, etc.). The toxicological and exposure threats are evaluated and the appropriate personal protective gear must be worn. The protective levels for workers are documented in the SARA 1986, 29 CFR 1910.120, NIOSH regulations and guidance documents for Levels A through D (Level A most restrictive and provides highest level of safety, and Level D for minimum safety level). Obviously, the work should not proceed until all the safety issues are resolved, and a highly trained and experienced safety officer should assist the hydrogeologist consultant in safety plan preparation.

Drilling contractor selection is probably the most important decision made to execute the subsurface exploration. The driller should be highly experienced with the drilling method and groundwater monitoring well construction, and

FIGURE 1 EXAMPLE OF BID SHEET

QUOTE NO: _____
BY: _____
PROJECT LOCATION: _____

TIME x RATE $

PRE-FIELD
Site Walk _____ hr. x _____
Research _____ hr. x _____
Meeting _____ hr. x _____
USA - site clearance _____ hr. x _____
Other (air photos, maps etc.) _____
Permits _____ hr. x _____   Subtotal _____

FIELD
Drilling _____ hr. x _____
  Prep to rig, cleaning _____ hr. x _____
  Steam clean per hole _____ hr. x _____
  Casing   Blank _____ ft. x _____
            Slot _____ ft. x _____
            Grout _____ x _____
  Finish Materials _____
Geologist - Log _____ hr. x _____
  Geo Technician _____ hr. x _____
  Sample Materials _____
  Travel - Drill Rig _____ hr. x _____
  Geologist _____ hr. x _____
  Vehicle _____ ml. x _____
Sampler _____ hr. x _____
  Well Development _____ hr. x _____
  Well Sampling _____ hr. x _____
  Sample Vehicle _____ day x _____
  Travel _____ hr. x _____
V. and H. Survey _____   Subtotal _____
Per Diem _____ day x _____

CHEMISTRY ANALYSIS
Soils _____ x _____ per test(s) x 15%
Waters _____ x _____ per test(s) x 15% Subtotal _____

REPORTING - Letter or Full Report
Geologist _____ hr x. _____
Senior Review _____ hr. x _____
Typing _____ hr. x _____
Reproduction _____ ea. x _____
Drafting _____ hr. x _____
Computer Time _____ hr. x _____   Subtotal _____

MEETING
Geologist _____ hr. x _____
Engineer _____ hr. x _____
Travel _____ ml. x _____   Subtotal _____
Per Diem
CONTINGENCY   Subtotals x 10%   Subtotal _____
                                            TOTAL _____

Figure 1   Example of bid sheet.

preferably have experience in the geographic region of your study. While drilling companies may be local to a region, the hydrogeologist must decide whether these companies are capable of performing the work. Statements of qualifications regarding the companies' abilities should be solicited with their cost estimate. Also, the drilling contractor may be requested to bring all materials and equipment needed for the project, and this should be clearly negotiated prior to starting work to prevent delays and work stoppage, especially at remote sites. Subsurface work can be highly difficult and subsurface conditions are changeable, so an estimate of subsurface conditions anticipated should be discussed with the driller. This means a realistic time schedule for drilling given the anticipated conditions. Since the driller selection is typically by bid, the lowest bidder usually will be awarded the contract. However, the lowest bidder may not be the best and while the financial factors are important, the most important factor is the drilling company's ability to get the job done on time and budget.

Lastly, the consultant and subcontractors may need to deal with insurance and bonding requirements. As the liability for subsurface and groundwater contamination work has risen in the past years, so has the need for insurance and bonding. This can be a paper-intensive and drawn-out process, and may involve negotiation with corporate principals. An awareness of these peripheral issues is necessary because they may impose some restrictions on work and subcontractors and affect time schedules for work task completions.

Decontamination Procedures

Decontamination of drilling equipment and sampling tools is vital to any groundwater contamination investigation. The purpose of decontamination is to clean the equipment of materials that could cause inadvertent contamination in the borehole or between boreholes. If these procedures are not used properly or inconsistently, the field data could be suspect (see Table 1). Decontamination cannot be overlooked in investigation planning, and the level of decontamination will depend on the site problem and contaminant. Also, the fluid and drilling muck may be hazardous waste depending upon content, so disposal costs are usually involved. A designated decontamination area should be selected and approved prior to starting work. The following suggested approach could include:

1. The decontamination area should be close to but not in the work area proper. There should be all-weather access and security if needed.
2. Utilities (power and clean water) should be available. Sometimes the only source of water may be a fire hydrant (which may be metered and billed to the project) or the driller may need to carry water in a tank truck if working at a remote site.
3. Cleaning and drilling fluids must be collected and properly disposed of after field work. The disposition of fluids may entail chemical testing to ascertain whether they are hazardous, and if so, must be disposed of under the pertinent regulations. This can become a significant cost and should be factored into the budget.
4. All drilling and well construction materials must be clean prior to use in each borehole. Materials stockpiled on site should be covered and stored away from the decontamination area. If there is any doubt, decontamination should be performed. Cleaned materials and tools should be placed on clean plastic and not in contact with pavements or the ground, especially around the work area.

Table 1 List of Selected Cleaning Solutions Used for Equipment Decontamination (Moberly, 1985)

Chemical	Solution	Uses/remarks
Clean portable water	None	Used under high pressure or steam to remove heavy mud, etc. or to rinse other solutions
Low-sudsing detergents (alconox)	Follow manufacturer's directions	General all-purpose cleaner
Sodium carbonate (washing soda)	4#/10 gal water	Effective for neutralizing organic acids, heavy metals, metal processing wastes
Sodium bicarbonate (baking soda)	4#/10 gal water	Used to neutralize either base or neutral acid contaminants
Trisodium phosphate (TSP oakite)	2#/10 gal water	Similar to sodium carbonate
	4#/10 gal water	Useful for solvents & organic compounds (such as toluene, chloroform, trichloroethylene), PBBs and PCBs
Calcium hydrochloride (HTH)	8#/10 gal water	Disinfectant, bleaching, & oxidizing agent used for pesticides, fungicides, chlorinated phenols, dioxins, cyanides, ammonia, & other nonacidic inorganic wastes
Hydrochloric acid	1 pt/10 gal water	Used for inorganic bases, alkali, and caustic wastes
Citric, tartaric, oxalic acids (or their respective salts)	4#/10 gal water	Used to clean heavy metal contamination
Organic solvents (acetone, methanol, methylene chloride)	Concentrated	Used to clean equipment contaminated with organics or well casing to remove surface oils, etc.

From U.S. EPA, 1989. With permission.

5. Typically two types of decontamination are used; hand washing and mechanical washing. Hand washing would be used for sampling tools, jars, sample liners, and other reusable equipment. Approved cleaners (usually low-sudsing phosphate, carbonate, or hypochlorite) and clean water are used. If rinses are used, they may be solvents or with deionized water depending on project needs. Mechanical washing is usually done with a steam cleaner that shoots heated water through a spray nozzle. This high-speed jet of hot water removes caked mud, drill cuttings, and volatile contaminants. It can quickly clean large pieces of gear and vehicles, and gets into crevices and tight areas. Cleaning is performed on plastic sheets or within bermed pavements and all fluid and mud collected in tubs or drums for storage and off-site disposal.

Rapid Site Reconnaissance Subsurface Sampling

At times the client may want to "get a feel" for the size of a problem quickly to ascertain whether to seek a more in-depth site study, or as a precursor to purchasing property, or to see if a plume is moving onto the site. In this case, the site subsurface study should be fast and inexpensive to get preliminary (or reconnaissance) data. Several approaches to reconnaissance site studies have been developed: geophysical (without invasive testing), soil vapor sampling, and shallow groundwater sampling (invasive testing). These techniques allow a quick look at a site to ascertain whether a contamination problem is present, and if so, to get a relative size estimate. The important thing to remember is that these are preliminary studies and often need additional sampling to verify the conclusions drawn from these limited data.

Geophysical (noninvasive) testing uses geophysics to ascertain the presence of buried bodies, to estimate depth to groundwater, and to get a gross handle on site geology. The review of all these techniques is beyond the scope of this book, but a quick review should introduce the reader to the concepts (see Sara, 1994). Magnetometers use magnetics to locate metallic objects (such as drums) or to determine a change in rock lithology, and by traversing the area, the investigator may get a rough location to center the limited subsurface effort (such as excavating). Magnetics may allow from several feet to several tens of feet of vertical penetration. A variation of this approach is to use imaging radar to get a view of the subsurface by the reflectance of the radar waves. The depth of radar penetration may be less than 20 feet. Electrical resistivity can be used to estimate depth to groundwater up to several tens of feet, or deeper, by sending electrical current into the ground. The amount of resistivity in the groundwater may be used to estimate groundwater depth and quality. Seismic refraction is used to determine reflecting horizons of denser sediment of rock with depth by sending acoustic signals into the ground and measuring the return time with geophones. This technique has been used for years to estimate the rippability of rock for engineering studies; it gives estimates of alluvial cover and possible rock type.

Soil vapor sampling is one of the "newer" reconnaissance techniques that uses the vapor phase sampling in soil to locate and map volatile contaminants. Volatile components of liquid contaminants will evolve vapor due to the contaminant vapor pressure. Vapor movement may be somewhat analogous to fluid flow in terms of gross porosity and permeability controls (see Figure 2).

Soil vapor mapping is accomplished by pushing a thin-wall, small-diameter tube into the subsurface (usually 5 to 10 feet, and up to 20 feet) using a one-way drive point (see Figure 3A). Once the desired depth is reached, the tube is pulled back slightly and the vapor in the tube is evacuated to induce soil vapor to fill the tube. A sample is drawn into a syringe, which is transferred to a portable gas analyzer, or into a laboratory sample bag for analysis at an analytical laboratory. Upon completion, the tube is withdrawn and the hole backfilled with a thin grout slurry.

Soil vapor surveys may require drilling permits, and a permitting fee may be required. Soil vapor surveys are rapid reconnaissance subsurface assessment techniques. Once the presence of vapors is analyzed and mapped, a contour map of vapor occurrence and concentration results. This map may be used to target possible sources, and to select monitoring well locations. However, the presence of vapors does not necessarily identify the source, and the technique is usable with volatile compounds. Semivolatile compounds may remain "unseen" and the technique is not really useful at all for trace elements of some pesticides.

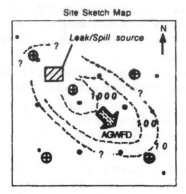

AGWFD- Assumed groundwater flow
direction from background study

•   Soil vapor point to locate possible
        plume location for "best guess"
        monitoring well location

⊕   Well location to provide coverage
        of plume with minimum number
        of wells

⌐--- 1 0 0   Soil vapor defined assumed
             plume position, contaminants
             in ppb

Figure 2   Initially located groundwater plume using soil vapor. The proposed monitoring
well locations are chosen on the basis of the source and the assumed location
of the dissolved constituents based on the vapor data. Only installing wells will
confirm the actual plume location.

Shallow groundwater sampling (see Figure 3B) is a variant of soil vapor
mapping and may be used in tandem with vapor mapping. A thin-wall tube is
advanced as discussed above, but it is pushed to the groundwater occurrence and
slightly into it. The tube may be periodically stopped for soil vapor readings and,
once advanced into water, to collect a sample. Although deep penetration can be
achieved, the sampling pipe is relatively light duty. This implies depth of "shallow" groundwater sampling at about 20 to 25 feet depending on sediment type.
A water sample is then analyzed and the "grab" sample results allow for a specific
piece of data to determine general conditions at that location. When used with a
mobile analytical laboratory, this can be a powerful rapid assessment technique,
especially in difficult locations, such as in roads or alleys or in buildings, and
where soil cuttings quantities are minimal for storage and disposal.

INTRODUCTION TO SUBSURFACE DRILLING AND SAMPLING

A number of drilling technologies are available for subsurface sampling and
exploration. The techniques may be designed for generic or highly specific drilling
conditions and depending on formation and type of work, some are better than

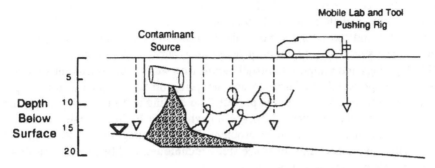

A. Conceptual Approach to Soil Vapor Sampling

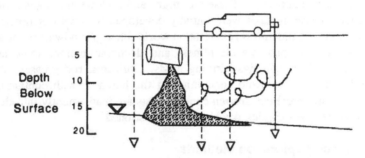

B. Conceptual Approach to Shallow Groundwater Sampling

Figure 3 Reconnaissance assessment by soil vapor mapping and shallow groundwater sampling. A. Soil vapor mapping. A probe is hydraulically pressed into the ground and a soil vapor sample is evacuated, then analyzed in the mobile laboratory. B. Shallow groundwater sampling. The shallow groundwater sampling approach is similar, only a groundwater sample is collected and analyzed either in the field or laboratory. This can be an effective rapid reconnaissance technique to ascertain presence of a problem or to locate permanent monitoring wells.

others. There is no universal drilling technique applicable in all subsurface conditions, and given time and budget constraints, previous subsurface work experience becomes highly valuable. The project hydrogeologists should have a strong working understanding of the drilling technique in order to stay within budget, especially if unforeseen problems occur. Driscoll (1986) presents an excellent review of drilling methods used in the water resources industry, and government guidance documents review methods used in groundwater contamination work. For a general review of subsurface exploration and sampling methods, see Leroy et al. (1977). The drilling soil or rock cuttings can be contaminated and could be classified as hazardous waste. Therefore, contingencies must be made to contain this material and properly store and dispose of it during and following the field work.

Sometimes drilling methods may be restricted due to site access, formation problems, and occasionally regulatory input. However, the ultimate selection of drilling equipment should always be based on the type of geology and investigative sampling needs. Subsurface drilling requires powerful, durable equipment and tools for the borehole diameter for the contaminant (or any) subsurface study. The exploratory borehole is advanced to obtain subsurface lithology logs and samples, to install monitoring wells. The consultant's knowledge and applicability of drilling and sampling directly relates to the budget and time schedule allotted to the project. The drilling and sampling provide the direct observations of the subsurface on which the hydrogeologist bases his opinions about the contaminant problem. People managing this work should be highly experienced in subsurface study techniques, and be able troubleshoot field problems, which will occur.

All drilling techniques are not reviewed in this text. The following discussion will broadly group three drilling techniques "most commonly" used for groundwater contamination work. These are: flight and hollowstem augers, rotary, and cable tool. These methods were initially developed for civil engineering, water resource, and oil and mineral exploration work. Since groundwater contamination work strongly emphasizes cleanliness and decontamination, some accessory materials used on drill rigs (drilling muds, lubricants, rod greases) should be carefully reviewed to avoid contaminating the borehole with contaminants for which you are searching. If such contaminants are inadvertently introduced into the borehole, the sampling data could be questioned.

Subsurface Exploration Methods

Sampling and lithologic data is collected by advancing exploratory boreholes into the subsurface. The borehole may be advanced to any depth, and into any type of formation or alluvial sediment. Again, drilling method choice depends on formation type, anticipated conditions, and ultimate depth. Subsurface conditions are changeable, and the driller and drilling method must be adaptable to compensate. Samples and lithologic logging are done while the borehole is advanced; upon completion, the hole is converted to a monitoring well or other instrument, or sealed to the surface.

Drilling methods are variations on a similar theme. The drill augers or rotary rods are manufactured in some convenient length (augers, 5-foot; rotary 5-, 10-, or 20-foot lengths). The drill bit grinds or scrapes the soil, sediment, or rock at the end of the first auger or rod. Upon reaching that length, another auger or rod length is attached at the surface and the drilling continues. When a sample is required, the drill "string" is hoisted to the surface and the sampling tools are lowered to sample ahead of the drill bit. Once the sampler is recovered, the drill string is again lowered with another length and drilling resumes. These steps repeat until the bottom of the the hole is reached, and the hole is then sealed, or a well is constructed. The following types of drilling are briefly discussed, but the drilling routine is very similar whether the depth is 10, 100, or 1000 feet. As the borehole becomes deeper, the lifting and replacing of drill rods or augers becomes longer for the "trip." The space in the borehole between the borehole wall and the auger fins or drill rod is called the annular space (see Driscoll, 1986; Keely and Boateng, 1987; LeRoy and LeRoy, 1977; Sara, 1994; University of Missouri, 1981).

Drilling Methods

Flight auger drilling is a common type of drilling technology because it is fast and economical to depths of about 100 feet in noncaving alluvium and soft rock. The helical auger bit cuts into the subsurface and the cuttings are lifted to the surface on the fins. Although auger drill rigs may be rated to depths of 150 to 250 feet drilling, the author's experience is that for depth below 100 feet they should be evaluated with the driller prior to advancing the hole. Augers are versatile, very durable, available in a wide range of diameters, and used in all states. Flight augers and hollowstem augers are used in many investigations (see Figures 4 through 6). These rigs and auger types are detailed by the drilling manufacturer and the aforementioned references. It is also prudent to contact the drilling equipment manufacturer regarding the uses and limitations of the equipment.

Flight Augers

Flight augers range from 3 to 36 inches or more in diameter, depending on the manufacturer, and are commonly used to drill boreholes from 1 to 100 feet deep. Augers must be withdrawn to access sampling tools, and thus augers are typically used in soil engineering and sampling studies. They are very good for groundwater occurrence and recognition because it is a "dry" drilling method. Borehole advancement problems may occur if the formation caves when augers are withdrawn. This tends to limit drilling deep holes, and when groundwater is encountered.

Hollowstem Augers

Hollowstem augers are light augers with a hollow core or stem whose inside diameter varies from roughly 3 to 8 inches depending on the manufacturer. Again, these augers are advanced in boreholes up to 150 feet and deeper depending on drill rig and geology. Since these augers effectively "case" the borehole as it advances, borehole collapse or caving is prevented. Sampling and monitoring tools can be sent through the hollowstem by removing the centerplug and rods (see Figure 4). This technique is excellent for water recognition and small diameter monitoring well installation. It can also be used with dry coring equipment to collect continuous core. Borehole problems may occur when hollowstem augers are advanced into saturated clean sand formations, where the sand may flow up into the augers from subsurface pressure, "locking" the augers into the ground. Auger removal is difficult because the borehole has effectively collapsed on the augers, and removal is difficult and time-consuming.

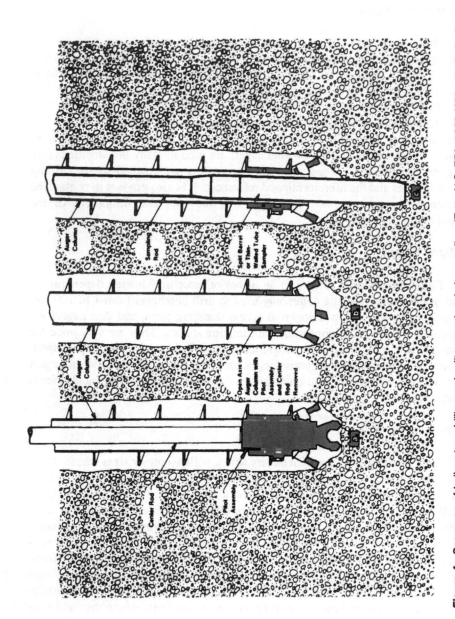

Figure 4 Sequence of hollowstem drilling and split spoon borehole sampling. (From U.S. EPA, 1989. With permission.)

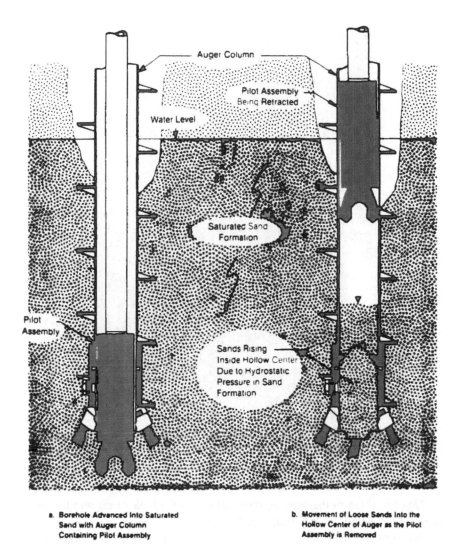

a. Borehole Advanced Into Saturated
Sand with Auger Column
Containing Pilot Assembly

b. Movement of Loose Sands Into the
Hollow Center of Auger as the Pilot
Assembly is Removed

Figure 5 Diagram showing hollowstem drilling into a heaving sand. (From U.S. EPA, 1989.
With permission.)

Rotary Drilling

Rotary drilling drills a variable diameter hole (typically 4- to 16-inch for
contaminant study work) from depths of tens of feet to hundreds or thousands
of feet; oil industry users may commonly drill boreholes that are miles deep. This
technique uses a fluid that is circulated from a tank or pit down through a hollow
drill rod and out through a drill bit. The rotary bit cuts or crushed the sediment
or rock and the cuttings are flushed up the outside of the drill rod in the borehole
to the surface. The weight of the drilling fluid (usually a mud–water mixture)
holds the borehole annular space open and lubricates and cools the bit. Once the

Figure 6 Photo of hollowstem drilling rig. (Photo by the author.)

cuttings are lifted to the surface, they settle into the tank or pit and the mud is recycled back into the drill rods (see Figure 7).

Samples are collected continuously by sieving cuttings as they leave the borehole at the surface, or conventional coring or interval sampling tools can be used at the desired sampling depth. This technique can be somewhat difficult to use in water recognition because the borehole is flooded. This method is used with other lithology logging tools (electric logs and tools using the drilling fluid, and in an uncased borehole). Variations of rotary drilling use air as the fluid and can be a quick method in stable rock or crystalline formations.

Contrary to some popular opinions, mud rotary drilling is appropriate for groundwater contamination work. The mud will seal the borehole and the water-bearing horizons effectively (Healy, 1989). Although some contaminants would enter and mix with the mud, the small quantity is acceptable and is removed when the borehole is bailed and flushed prior to sealing or well installation. When experience and qualified drillers are employed, this method is fast, economical, and adaptable to changeable subsurface conditions. Generally, if the boreholes

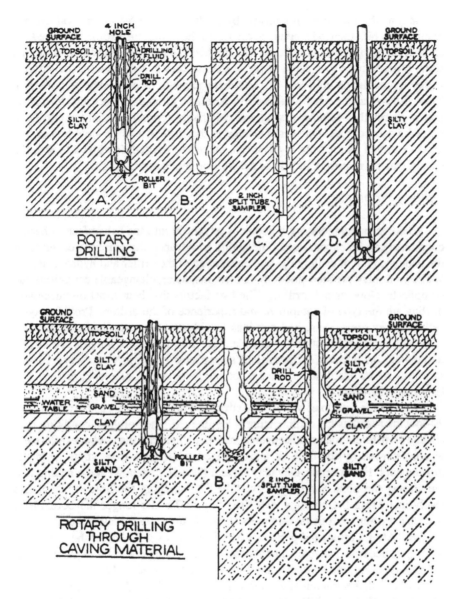

Figure 7 Mud rotary drilling. (Modified from University of Missouri, Rolla, Seminar for Drillers and Exploration Managers, 1981. With permission.)

are to be advanced to depths of 150 feet or deeper, rotary drilling use should be considered.

Cable Tool Drilling

The third technique is cable tool drilling, which is also the oldest method of drilling (Driscoll, 1986). This method drills a variable diameter borehole (6 to

36 inch) to shallow to deep intervals by drilling a 4- to 5-foot interval using a hammering bit, then advancing a casing to hold the formation, followed by another drilled interval and casing. The method is essentially fluid free, although some water is used to make a slurry of the cuttings to bail them to the surface. Either a continuous chip log, discrete drive sample, or core may be collected. This method can be used to advance a borehole through almost any formation and is especially useful in caving and collapsing formations that contain large boulders and cobbles. One disadvantage is the method can be slow and casing must be driven as the hole is advanced. Once the desired depth is attained, the casing may be removed, or left as the well or conductor casing.

Drilling Problems

Whenever an exploratory borehole is advanced into the subsurface, different conditions can be encountered. The ability to advance the borehole depends on the geology, type of drilling rig, and experience of the driller and hydrogeologist. While work may proceed smoothly in most instances, changeable conditions can complicate, slow, or halt drilling. The two factors that bear most on successful drilling are the type of equipment and experience of the drillers. Problems typically arise at some point, and previous troubleshooting experience of the drill crew and hydrogeologist are invaluable. The drilling budget may be the largest dollar item on the cost proposal and severe cost overruns could occur; some problems are avoidable and some are not, but recognizing incipient problems can help minimize them.

Exploratory boreholes are usually advanced in increments (usually 5-, 10- or 20-foot intervals) as the drilling stem is lengthened. The borehole condition may change as different material is encountered. Each material may behave differently when disturbed by the drill bit and hydrated with water from strata or in the drilling fluid. For example, clay may expand ("swell") when wetted pinching into the closing the hole. Sand may cave ("bell"), enlarging the hole diameter and vastly increasing the cuttings yield (see Figure 7). The longer the borehole is open and drilling tools travel the borehole length, the more prevalent the "borehole erosion." After a prolonged period (several hours to over 24 hours) the borehole's walls may become unstable and collapse, and redrilling could be required. If the collapse is severe, the borehole may have to be abandoned and a new hole drilled next to the first. A variation of this problem is drilling through cobble and boulder-sized rocks where the drill bit cannot "bite" and crush the particles. If bouldery conditions exist, then that zone may be sealed by installing a conductor pipe into and through it. Hence, this condition could require first using cable tool, then augers or rotary.

Another problem is flowing sand. A flowing sand condition occurs where sand is relatively of uniform diameters, is saturated, and contains very little silt or clay. The action of the rotating drill bit causes the sand to liquify and "flow" into the borehole. When this happens using hollowstem augers, the sand can flow up into the augers, plugging and locking them into the ground. When rotary mud is used, the sand may flow and "grab" the drill rod, halting bit advance. When severe flowing sands occur, a possible solution is to drill slightly deeper with the augers to attempt to find a stratum that does not flow so the augers can then be lifted to the surface and cleared. When mud rotary drilling, a possible solution is thickening the mud to a heavier weight, which can stabilize and seal the borehole wall and annular space open.

Finally, each borehole may have some amount of caving and slough from the borehole excavation at the bottom. If severe caving conditions are a problem, the ultimate well completion may not be as deep as originally designed. A separate borehole drilled to the design depth may have to advance following the first in which the samples and lithology data were collected, and within several feet of the first, for the well. These are field judgments that may need to be made by the driller, hydrogeologist, and office supervisor. The acceptable amount of drilling slough, borehole depth, caving and erosion, modifications to well design, sampling locations and depths, and so on, should be discussed prior to going into the field. This can save money and time since coordinating by telephone with office people in different cities or states and over weekends or early morning or night is almost always a coordination dilemma.

SOIL SAMPLING METHODS

Driven Samplers

Soil sampling methods have evolved over the last 30 to 40 years for consulting use, and have been used extensively in civil engineering (note that soil means soil, unconsolidated sediment, or weakly consolidated sediment). Samplers and sampling tools used on both auger and rotary rigs use a driven sampling system where a hollow steel tube and shoe cutting bit, sometimes called the "spoon," is advanced by hammer blows ahead of the drill bit. Several variations of this sampler type exist. The standard penetrometer is a 1.7-inch-diameter, 18-inch-long sampler that collects soil samples allowing direct observation of soil stratigraphy and allowing field engineering measurements to be taken. The samplers are then split longitudinally and the soil "core" may be observed (see Figure 8).

The number of hammer blows on the sampler is recorded and can yield indigenous strength characteristics of the soil. The drive method uses a 140-pound hammer dropped over a 30-inch fall. Driven samplers are all driven in the same manner: The sampler is lowered at the desired depth interval through the hollow-stem or in the open borehole. A portion of the drill rod attached to the sampler is marked into three 6-inch increments (for the 18-inch-long sampler and drive bit). The hammer is then used to drive the sampler into the ground, and the number of blows needed for each 6-inch drive is recorded. The actual drive "blow count" is the last two 6-inch (the initial 6-inch drive is through the drill slough to get below the bit, which can be 2 to 6 inches long). The blow count is recorded on the log, the sampler is hoisted to the surface and disassembled, and the sample is lithologically logged.

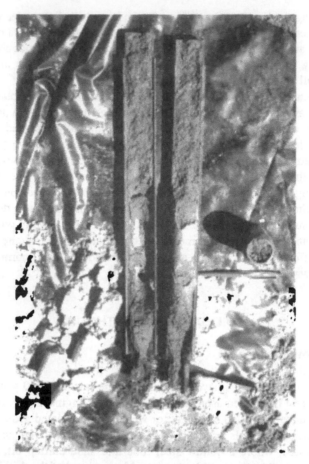

Figure 8   A standard penetrometer, split to show soil texture; pen points to contact of upper
            saturated sand and lower damp clay. (Photo by the author.)

These samples are collected at any depth interval, typically at the end of
each drilling auger or rod. Hence the logger looks at about 1 foot of stratigraphy
every 5 feet in boreholes. The intervals are repeated as the hole is advanced,
allowing a relatively good look at the subsurface in a quick and cost-effective
manner.

Variations on this technique include the California modified and California
split spoon samplers. The California modified sampler is a 2.0-inch inside diameter, 18-inch-long steel tube into which sampler liners can be placed to collect
the soil. The split spoon sampler is similar to the modified sampler except the
inside diameter is 2.37 inches and more sampler rings can be inserted. These
sampler rings allow for the collection of a "relatively undisturbed" sample (it
must be slightly disturbed for collection), returned for laboratory strength testing.
Contaminant hydrogeology work allows these rings' ends to be sealed for chemical analysis with minimal outgassing, allowing versatile use.

Other Sampling Tools

Additional sampling techniques are available to collect larger soil samples or continuous samples, depending on the need. The piston, or Shelby Tube sampler, hydraulically pushes a 3-inch inside diameter tube into soil ahead of the bit. The tube is then retrieved, sealed, and sent to the laboratory for geotechnical soil and chemical tests. This is a common sampling technique for large civil engineering projects.

Soil auger coring is similar to conventional hardrock coring. A core is similar to the soil core described above; however, at times a continuous core is desired. The continuous core allows observation of the stratigraphy for fine detail in the lithologic log to describe very small features on interest. The coring barrel is a steel tube (usually 5 or 10 feet long) with a cutting bit. The core is lowered into the hollowstem or into the mud hole and rotated at high speed, and the borehole is advanced. However, the core is collected in the barrel rather than as cuttings. When the core run is finished, the barrel is hoisted to the surface, the core barrel is disassembled, and the core is placed in a core box for logging and storage. Coring is usually more expensive than interval sampling, so care is taken to recover, package, and store as much as possible as a site geology reference.

EXPLORATORY BOREHOLE LOGGING

Accurate borehole logging is essential for the success of any subsurface investigation. The logging technique related to groundwater studies is primarily a lithologic log of the soil sediment rock encountered. It also includes the occurrence of groundwater, depth and speed of the drill bit advance, drilling method and equipment, and relevant site data at the drilling site. This log contains the primary observed information of the aquifer and aquitard, sampling depths, and pertinent geologic observations and rationale for monitoring well construction. If the lithologic logging is incomplete or incorrect, subsurface conditions cannot be understood and valuable data is lost. Worse yet, the interpretations to other boreholes and sampling locations can lead to very erroneous conclusions and severely compromise the project.

Because subsurface data collection techniques were developed in civil engineering work, the logging style is a hybrid of both soil engineering and geologic logging. The hydrogeologic emphasis is needed for the obvious reasons, although the soil engineering aspects are used nationwide and must be understood by the person doing the logging. Since many investigations deal with surficial soil and alluvial deposits, the following discussion is primarily aimed at soil alluvium and sediment logging.

The Unified Soil Classification System (USCS; see Casagrande, 1948) is almost universally used for all soil logging and engineering geologic work (see Figure 9). The method is contained in U. S. Army Corps of Engineers guidance manuals. The system is presented in the American Society of Testing Methods

Test 1 COARSE-GRAINED SOILS — More than half of the material (by weight) is individual grains visible to the naked eye.	Test 2a GRAVELLY SOILS—More than half of coarse fraction is larger than 1/4".	Test 2b CLEAN GRAVELS Will not leave a stain on a wet palm	Test 2c: Substantial amounts of all grain particle sizes			GW
			Predominantly one size or a range of sizes with intermediate sizes missing			GP
		DIRTY GRAVELS Will leave a stain on a wet palm	Test 4: Nonplastic fines			GM
			Plastic fines			GC
	Test 2a SANDY SOILS—More than half of coarse fraction is smaller than 1/4".	Test 2b CLEAN SANDS Will not leave a stain on a wet palm	Test 2c: Wide range in grain size and substantial amounts of all grain particle sizes			SW
			Predominantly one size or a range of sizes with intermediate sizes missing			SP
		DIRTY SANDS Will leave a stain on a wet palm	Test 4: Nonplastic fines			SM
			Plastic fines			SC

Test 3 FINE-GRAINED SOILS More than half of the material (by weight) is individual grains not visible to the naked eye.	Test 5—RIBBON	Test 6—LIQUID LIMIT	Test 7—DRY CRUSHING STRENGTH	Test 8—DILATANCY REACTION	Test 9 TOUGHNESS	Test 10 STICKINESS	
	None	<50	None to Slight	Rapid	Low	None	ML
	Weak	<50	Medium to High	None to Very Slow	Medium to High	Medium	CL
	Strong	>50	Slight to Medium	Slow to None	Medium	Low	MH
	Very Strong	>50	High to Very High	None	High	Very High	CH

Test 11—HIGHLY ORGANIC SOILS	Readily identified by color, odor, spongy feel, and frequently by fibrous texture.	OL
		OH
		Pt

Figure 9   Summary of the Unified Soil Classification System. (From U.S. EPA, 1991b. With permission.)

(ASTM) D-2487 and 2488 and provides the textural classification and engineering performance of gravel, sand, silt, and clay (ASTM, 1988). These textures are similar to geologic texture classifications, and are based on standard U. S. standard sieve sizes. The USCS classifies on the basis of soil texture, engineering performance related to moisture content, plasticity, and strength. While these data may be used in some aspects of work, the primary emphasis here is on describing subsurface groundwater, geology, and contaminants. Additional information is also logged relating to primary geologic features, biologic features, and contaminant staining.

The geologic data must supplement the purely textural USCS data for the log's usefulness to both engineers and hydrogeologists. The hydrogeologic information will provide direct insight into the hydrogeology, especially the porosity and permeability of soil or sediment. This should include at the minimum the following: bedding thickness description, sorting (opposite of grading), presence of biologic structures (rootholes, burrows, their interconnection and orientation to bedding), and contact relations between strata. These data are vital to observe and interpret the following: grain-to-grain packing and presence of fine material that clogs porosity, indigenous structures which may allow crosscutting openings for preferential contaminant movement, and spatial relationships of individual strata. Bedding may yield insight into preferential flow along horizontal bedding, or possible vertical penetration in the absence of bedding. Obviously, the nature and relationship of the aquifers and aquitards is the prime interest, and the investigation becomes almost meaningless without data collection with hydrogeologic emphasis.

Logging aids are available that the geologist, hydrogeologist, or engineer should always use to increase accuracy. These include a hand lens for close observations, color references (Munsell Color for soil; Geological Society of America Rock Color), grain size keys and percentage aids, see-through ruler, and a protective notebook. An aluminum notebook is very handy since you are also carrying drilling permits, maps, other logs, safety information, and other documents; water and mud can obscure or destroy paperwork, so it is wise to protect it.

Field soil vapor screening is common in today's field work, so a portable volatile organic analyzer may be used. This can perform simple headspace tests of outgassing soil in plastic bags or glass jars. Another approach is a field colorimetric test kit, which gives a semiqualitative concentration of the contaminant. As a general rule, the portable analyzers can detect volatile materials (depending on the detection device) such as gasoline and solvents. Field colorimetric kits may be used to estimate volatile as well as less volatile content in soil. This data is logged onto the boring log at the sample interval. It is very important to remember that the field screening and detection of gas is not necessarily proof of the contaminant presence in the soil matrix (confirmation should always come from rigorous laboratory chemical analysis).

Logging Procedure

A suggested logging procedure is presented that could be used to log boreholes for the groundwater contamination purpose. Since field work is expensive, and the information is vital to the project's success, it is the responsibility of the geologist, hydrogeologist, or engineer to properly log and collect all the information from each borehole. Thus, a system should be utilized that can collect the data to answer the questions for that site, which is the primary purpose of the hydrogeologist's log (see Figure 10).

The soil lithology logging procedure would be as follows:

1. Drive the sampler at the desired depth interval. Record the blow count (or push hydraulic pressure if a push sampler is used).
2. Remove the sample from the sampler in the correct orientation (typically the top of the sample is shallow and the bottom is deep). Package and seal any samples retained for chemical analysis. This is needed at this point to prevent volatilization or inadvertent contamination of the sample. The packaging used would be appropriate for the suspected contaminant.
3. Lithologically log the soil or sediment. A suggestion for completeness is always to collect the data in the same order; the point of the order is to make information collection almost automatic, but it should not be used as a shortcut, or not collecting additional data at a depth of interest. The data collected would include: Texture (USCS-ASTM classification); Color (Munsell Color Chart); Percentage of Gravel, Sand, Silt, and Clay; Plasticity of Silt and Clay; Presence of Bedding; Presence of Biologic Structures; Field Evidence of Contamination (staining, odor); Consistency and/or Density (from the hammer blow count); Relative Moisture Content (i.e., damp, moist, saturated — yields free water); and any other features of interest.
4. Other notes. The drilling rate should be logged and the presence of drilling difficulty or ease should be noted. This may indicate strata that are difficult to sample or an upcoming change in texture and strata. The first occurrence of groundwater MUST be logged and the depth should be measured as accurately as possible.

One should remember that the textural classification is for soil and not rock (rock logging will be introduced below). Lithified sediment may display logging characteristics related to USCS in the field, but degree of lithification (a sedimentary rock-forming process) should be noted on the log. It is the author's opinion based on field experience that the aforementioned approach should yield the minimum information without sacrificing accuracy, yet collect data expediently. Each individual typically develops a routine as experience is gained and should use a system that gathers all the data yet is easy and comfortable to use.

Rock Logging

The approach to rock logging is conceptually similar to the soil/sediment logging (see Sara, 1994; U. S. Department of the Interior 1990; Williamson, undated).

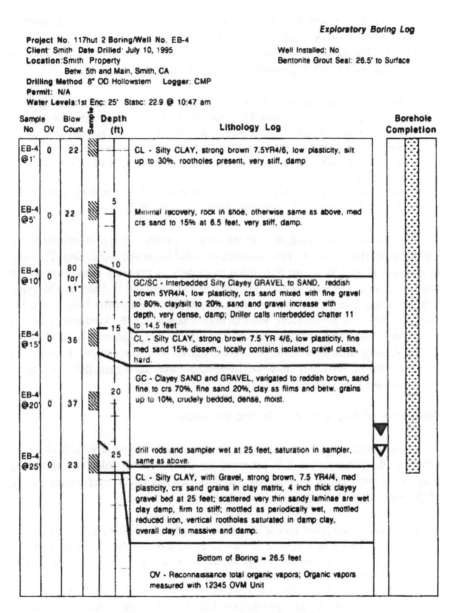

Figure 10 Example of an exploratory boring log.

Again, one collects the pertinent sample description information regarding the aquifers and aquitards and for contaminants in rock bodies. Hence, the log should contain the information relating to consolidated, indurated, or crystalline. The formation name and the contact relations of the overlying and underlying formation or stratigraphic unit should appear on the log. Several logging styles were developed for petroleum, mining, and engineering geologic needs. It is possible that if one logs rock, the drilling method may be rotary or cable tool. Thus, a continuous chip log with drive interval or coring sampling, together with electric logging tools, would be used in these boreholes.

The rock log would be similar in set up to the soil log with similar information goals. A suggested rock log format could the following:

1. Rock name, with formation name and age; lithologic descriptors, such as igneous, sedimentary, or metamorphic grain-size relations and properties of fresh and weathered surfaces; bedding, foliation or flow texture; weathering or alteration; hardness; fossils; discontinuities such as shears, fractures, and contacts, their attitude, spacing, infilling, density, length; rock quality designation; occurrence of water.
2. Log of drill bit penetration of pressure, percentage of core recovered, loss of drilling fluid, types of bits or core barrels, rods and drill rig used.

When continuous coring, usually the core is washed when it is brought to the surface and needs to be carefully handled to avoid breakage and drying. The core is typically placed in a core box, fitted as closely as possible to the orientation when retrieved. The core boxes must be carefully marked and stacked to avoid confusion and breakage. Often the core is photographed in the field, since the core will dry and some color aspects may change. Lastly, continuous coring sometimes means that one does not recover all the core. Core can be washed out by the drilling action, or slide out of the barrel. The percentage of each core interval actually retrieved must be logged, as well as where difficult coring conditions are encountered.

Detection of Groundwater in the Borehole

The detection of groundwater is not always clear and typically the driller may advance into water before it is observed in a collected sample. The following suggestions may aid in logging the first groundwater occurrence, which is one of the most vital observations in the borehole.

1. Always collect a soil sample if water is thought to occur at a soil interval of interest. This should be done so the hydrogeologist sees the occurrence in the soil, and extra samples should be driven to be sure if needed. This includes observing water marks on the drill rod.
2. Changes in drilling resistance or penetration rates can be useful. The rate of penetration may change if saturated conditions are encountered. Rates of drilling resistance should always be logged. A similar observation may be made with blow counts if the count varies in similar texture material, where increasing moisture affects soil strength (i.e, total blow counts decrease).
3. Increasing changes in observed soil moisture in samples with depth may infer approach of the capillary fringe.
4. Once drilling in a saturated unit, a sample should be taken to see if the drill bit has advanced into a "dry" unit (aquitard under the first aquifer). This is extremely important to prevent cross connection of aquifers during well construction, or to observe and note so that interconnection is minimized during the drilling. The seal through the aquitard needs to be written on the log.

5. The geologist should talk to the drillers and discuss what he anticipates in the course of the project, and any special depths of interest into the borehole. The drillers should be experienced with many and varied drilling conditions. The drillers' "feel" to the operation is very important, especially if they have worked frequently in the area.

6. Once water is encountered, water entry may be slow and the drilling may need to stop to ascertain that water is present. This is very important in drilling through low-permeability situations. While the duration of waiting may vary, the author uses 30 minutes if all field data indicate water could be present.

7. Groundwater depth measurements (including time and date) should be taken when water is encountered and compared to later depth measurements at the end of drilling the borehole. This gives an indication of the degree of confinement which could affect placement of well screens. A rule of thumb is that the faster and higher the water rises in the borehole, the more the confinement; if the rise exceeds 3 to 5 feet from the initially encountered groundwater depth, confinement is assumed for the borehole.

8. When rotary mud drilling, the mud may be observed to thin when groundwater is encountered. When air rotary drilling, moist air and damp cuttings are returned to the surface.

9. Downhole testing may be done in either soil or rock situations. A packer test may be used to seal a stratum of interest and either pump water out, or pump (clean) water into the formation. These can be especially useful in fractured formations, allowing rough aquifer yield estimates of the formation.

Subsurface information collection is the heart of any investigation because all the subsequent interpretations will be based upon the log and observations compiled in the field. Consequently, the interpretations are only as accurate as the observed data collected. Hydrogeologists, engineers, and geologists should be highly experienced in borehole logging and recognition of significant geologic features. This means logging numerous boreholes to increase the individual's ability. Each borehole is different; new insight is gained from logging experience and assists field personnel when they are supervised from the office. As in other endeavors, there is no substitute for experience in the field.

Direct Cone Penetration Testing

This brief review of cone penetration testing is presented because this is being used increasingly in site assessment and can yield valuable information, especially when used with the aforementioned borehole logging. The direct cone penetration testing (CPT) system is a static probing type of equipment that has been used in Europe since 1917 and in the U. S. since 1965 (Sara, 1994). This differs somewhat from conventional drilling in that a continuous reading of resistance is gathered from the borehole advancement. Different systems are used, but the basic approach is to push into the ground a rod equipped with a measuring tip. The tip measures resistance by a mechanical, hydraulic, or electric system (see DeRuiter, 1982; Sara, 1994; University of Missouri, 1981). The data collected is a continuous representation of the strata encountered and may yield very fine detail on strata thickness and internal variation. This helps in interpolation between logged boreholes and in correlation of strata. The log is a transcription of the data onto a paper roll somewhat similar to a geophysical log. In the recent past, some systems have been modified to collect soil and groundwater samples for contaminant work. However, the CPT has limitations in that a direct view of cuttings and samples is not necessarily present and the cone cannot penetrate dense sand and gravel, cobbles and boulders, and decomposed rock.

Borehole Geophysical Logging

Geophysical logging techniques (see Figure 11) have been used for many years in mining and oil exploration work, and in groundwater investigations. Conventional electric logging has been performed for years in water resources studies, and the same types of tools and logs can be used in contamination work. Usually the geophysical logging is done after collecting the soil or rock samples and aids in filling in the gaps between intervals of sampling and in revealing fine detail of the strata. Thus, electric logs are more commonly used in deep boreholes (over 100 feet) and in complex or unstable formations. A review of geophysical logs is beyond this book's scope, but a brief summary follows (for a more detailed review, see Keys and MacCary, 1971).

Geophysical logs are done in the open uncased borehole, and some logs (such as resistivity) require flooding the borehole. Geophysical logs that have general use in the groundwater studies are:

Resistivity — used to identify porous (sandy) sediment.
Spontaneous Potential — used to locate clay and sand strata.
Natural Gamma — used for lithology and to distinguish sand and clay strata.
Caliper — used to measure the borehole circumference and locate areas of erosion.
Drift — used to measure the inclination and position of the borehole bottom.

Usually logs are run together, although each tool must still be lowered into the borehole for its logging. When compared to the borehole log, it is an excellent check on the visual log and may help to correlate thin strata and identify contacts of interest. Interpretation of electric logs requires individual experience as mentioned above, as well as using an experienced geophysical logging contractor.

SUBSURFACE MAPPING AND STRATA CORRELATION

The exploration of the site with boreholes allows one to observe the soil, sediment, or rock, as well as the vertical and horizontal changes therein. As more boreholes are drilled, the boring logs can be used to create a working subsurface model as the work proceeds. This may be useful if boreholes are moved during the course of the study, or unanticipated conditions are encountered. Consequently, an ability to accurately log the geology and conceptualize while in the field aids in the understanding the geology, hydrogeology, and contaminant distribution. It allows the detail to be observed that may later become homogenized in assumptions

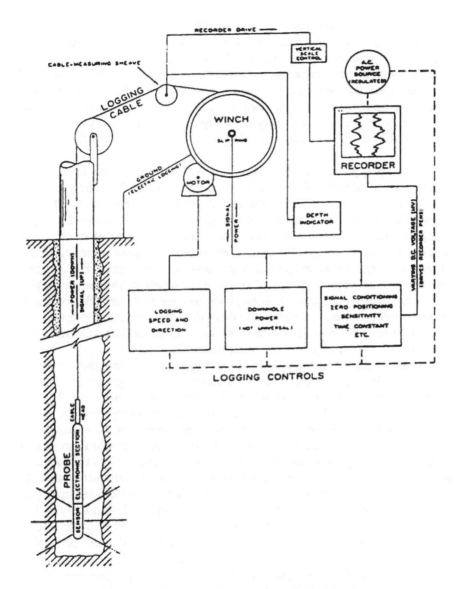

Figure 11 Schematic diagram of geophysical logging.

for preparing cross sections and other groundwater models. Unless there is validity for the assumptions, the ultimate site conceptualization and characterization could lead to erroneous conclusions. A very important goal is to correlate strata and predict contaminant occurrence and strata in which the contaminant may migrate.

Subsurface mapping allows one to construct a model of the subsurface using the relatively few boreholes to build a three-dimensional site model. This should yield a comprehensible view of the subsurface strata and aquifer and aquitard relationships for a proposed contamination and hydrogeologic definition for a remedial action. Subsurface mapping involves the identification of different types of strata and their extent. The identity of stratigraphic bodies is their texture.

These bodies are then correlated by their identifying characteristics areally across the site. The correlation is made from borehole to borehole by extrapolating the data. An understanding of geology and sedimentation and sedimentation environments is required to comprehend subsurface mapping and make meaningful interpretations. By recognizing these environments, and based on what you observe in the boreholes, you can predict the lateral change. This environmental reconstruction has been used widely in geologic study and can be used to reconstruct subsurface environments at any depth, scale, and time (see discussion in Chapter 1).

Stratigraphic principles can be used to make these correlations anywhere on earth. Strata are deposited from bottom to top, so obviously the underlying strata is older than than above it. Strata positions depend on the environment of deposition. That is, the vertical and lateral changes are environment-specific; if you recognize an environment, it will help you to understand whether strata will or will not be present. The stacking of sediment in time and space are repetitive and allow both interpretation and prediction of geologic environmental changes on a site at any scale. Figures 12 and 13 shows depositional environment examples where the sediment changes are noted given that type of deposit. By looking at these vertical and lateral changes, and where these sediments are observed, it is possible to build the geologic model. Note also that the sediment (gravel, sand, silt, and clay) can be logged easily using the USCS.

Figures 14 and 15 show borehole data examples and possible interpretations. These examples show samples collected at intervals and changes observed by the driller. This in turn can be used to interpret first the stratigraphy, which forms the geologic model. Then the geologic model can be interpreted into a site hydrostratigraphy. The hydrostratigraphy interpretation is a "homogenizing" step in that some more generalizations are made for the sequence that is an aquifer and aquitard. A range of sediment textures is present, and as a result the hydraulic conductivities may vary. The important point is that in each interpretive step, the geologic begins to "fuzz out" and the hydrogeologic model becomes more predominant. However, when numbers (i. e., porosity, hydraulic conductivity, bed thickness) are assigned to the model, the generalizing effect may increase or decrease water flow by over- or underestimating the porosity and permeability/hydraulic conductivity. Small errors in assumption may grow unless the "sanity checks" of geology and relevance to the hydrogeologic assumptions and interpretations are reviewed at each step.

Comments on Extent of Stratigraphic Units, Microstratigraphy, and Limits of Borehole Data

The lateral extent of stratigraphic units is a very important interpretation in all subsurface study. First the correct depositional model must be recognized. One may assume that the strata's properties are also laterally similar, which may or may not be true. The strata may pinch and swell in thickness, and lateral textural changes have obvious effects on estimates of porosity and related conductivity. Strata are not vertically monotonous, but may contain fine internal

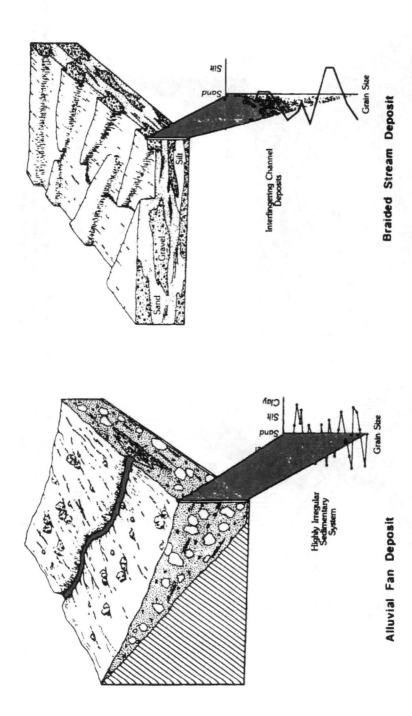

Figure 12   Depositional environment and sediment variability. A. Alluvial fan deposit. Little sorting of sediment, wide range of particle sizes of a borehole drilled through the deposit. B. Braided stream deposit. Gravel and sand are the predominant particle sizes and occur in sorted strata with little silt. Boreholes drilled into alluvial fans reveal few laterally extensive strata; braided streams show interbedded sands and gravels with little clay or silt. (From Mathewson, 1981. With permission.)

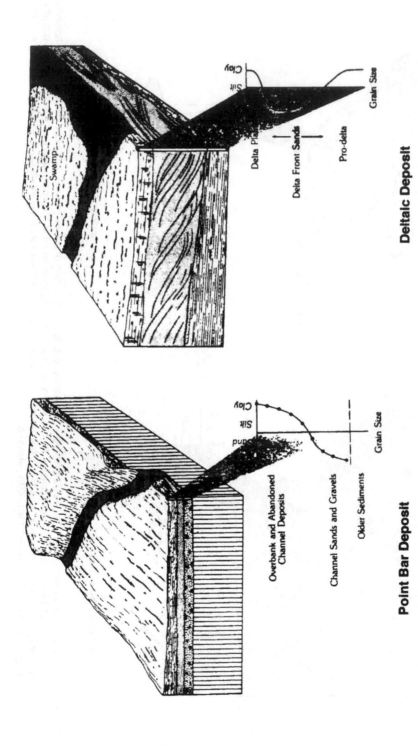

Figure 13   Depositional environment and sediment variability. A. Point bar deposit. Upward grading of gravel, sand, silt, and clay with cyclic strata vertically and laterally continuous units. B. Deltaic deposit. Laterally extensive sands bounded by thick clay below, and silt and clay above. Boreholes drilled in these depostis should reveal strata lateral continuity with laterally extensive clays. (From Mathewson, 1981. With permission.)

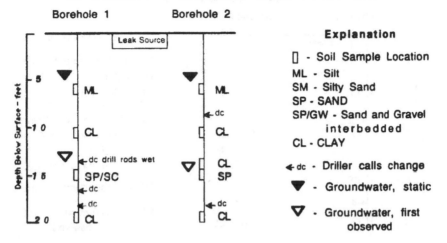

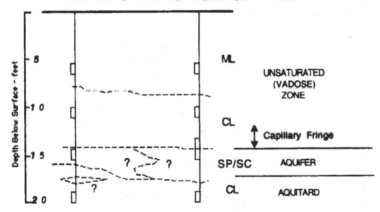

Figure 14   Hydrostratigraphy of simple confined aquifer, delineation for a site with two exploratory boreholes. Note the geologic stratigraphy in A and interpreted hydrostratigraphy in B. Aquifer is sand and clayey sand, so a range of hydraulic conductivities are present as well as interbeds. Roughly 10 feet of pressure head is observed in the boreholes.

stratification that may affect horizontal and vertical contaminant movement. Kueper et al. (1993) have shown in column experiments that internal thin stratification may affect the movement of tetrachloroethylene and its ultimate dispersal in sediment.

Many times internal stratification is less than a foot thick and these strata do not usually have widespread lateral continuity. Consequently, the correlation of these thin or discontinuous units over very long distances may be unreasonable unless observed in other boreholes. This may have some effect on prediction of contaminant movement. When a borehole is sampled on intervals, the distance between samples is not observed directly. One possible solution is to stagger the

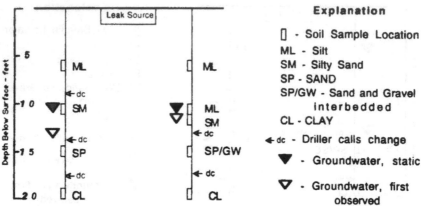

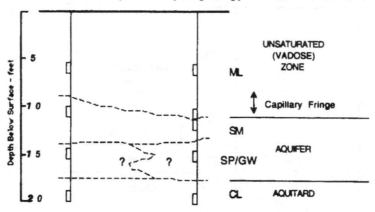

Figure 15   Aquifer delineation and hydrostratigraphy. Simple unconfined aquifer delineation for a site with two exploratory boreholes. Note the geologic stratigraphy in A and interpreted hydrostratigraphy in B; note range of possible hydraulic conductivities.

sample depths from borehole to borehole (see Figure 16). In this way, the depth interval tends to be staggered from hole to hole (that is, 10 to 11.5 feet in one hole, and 8.5 to 10 feet in the next) and is cost-effective. Obviously this is only a partial solution, and a continuous core is the remedy for observing the entire depth interval. However, a continuous core may reveal strata at thicknesses and textures that may vary laterally, so the next core may reveal other variations. A continuously cored borehole allows more observations, but will not yield all possible stratigraphic variations. As mentioned above, the field judgment may be to collect more interval samples or locally core sections to get the desired infor-

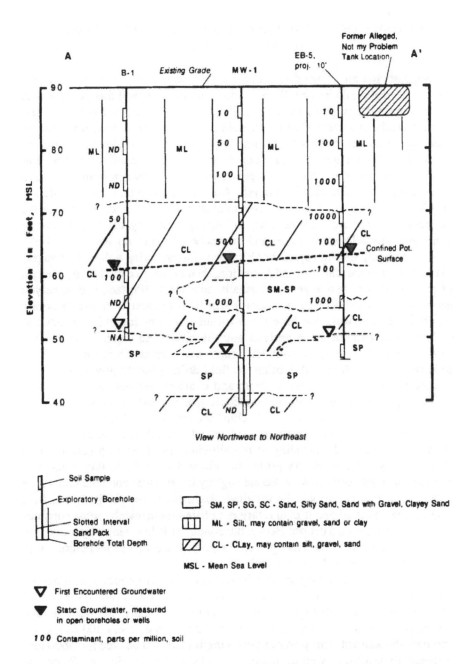

Figure 16 Example portraying geologic cross section and vadose contamination.

mation. Boreholes drilled outward from the continuous core would verify contacts of the major strata and correlation of strata of certain thicknesses, so that together with the depositional environment or geologic model, the site conditions may reasonably be constructed and lateral variations anticipated.

EXAMPLE OF INTEGRATING SAMPLING AND DRILLING IN
A DEEP BOREHOLE

Suppose you are retained to drill an exploratory boring 650 feet deep and install the smallest diameter fully penetrating monitoring well that you consider feasible given the available budget resources. The site geology consists of alluvium to depths of about 100 feet, underlain by a weakly consolidated formation. You have to collect samples of the alluvium for chemical analysis but the contamination does not appear to have penetrated into the formation based on previous information. You want to get a good idea of the formation stratigraphy and retain samples for permeability analysis. Also, an aquitard unit occurs at 450 to 600 feet and a pumping aquifer underlies the aquitard. Finally, the budget for the project is not large and while accuracy in logging is primary, the project must be done within the allotted resources and time schedule. How do you proceed?

It is probably wise to approach this in two phases; boring hollowstem augers to the contact of the alluvium and formation, then switch to a mud rotary method of drilling. Hollowstem augers will usually advance to 100 feet, and since you need to go another 550 feet, mud rotary will be the fastest with sample coring capability. In this way, you have the nonfluid soil sampling ability needed to collect samples for chemical analysis and the power and depth of penetration rotary affords. The weakly consolidated formation may have a tendency to cave or have severe borehole wall erosion, so fluid drilling should give you the best method to advance to the desired depth and complete the well. Also since you are drilling a deep hole you budgeted electric logs so sand and clay beds can be located and to supplement the visual "chip" log (see Figure 17).

The hollowstem auger borehole is advanced and soil samples are collected at the desired intervals including at the alluvium and formation contact. This should give you the chemistry profile the client desires. Now, this auger hole provides the pilot borehole for a second slightly larger auger borehole into which a steel conductor casing will be installed. Remember, the alluvium is at least slightly contaminated and if mud rotary is initiated immediately, some contaminants may be circulated deeper. Hence, the conductor is lowered and pressed into the formation and cemented into place. The grout is allowed to set before rotary drilling starts so the seal is not disturbed.

Now we start mud rotary drilling. This drilling yields a continuous lithology "chip" log so formation conditions can be viewed at all times. While the continuous chip log shows lithology, it is wise to core episodically, so given the time and money (coring is expensive) we will core 20 intervals every 100 feet as we approach the aquitard. This provides your samples that will be used for laboratory permeability testing, as well as lithology checks on the electric logs. Since you are aware that an aquitard is present at 450 feet, you should core an interval from 440 to 460 to verify the upper contact and allow you to view the contact relationships. Aquitard location is critical for pumping aquifer protection, so additional core samples will be collected from 510 to 530 and 580 to 620 to view the internal stratigraphy and the aquifer–aquitard contact. The monitoring well design depth is 650 feet, so the borehole is advanced to that depth. The aquifer

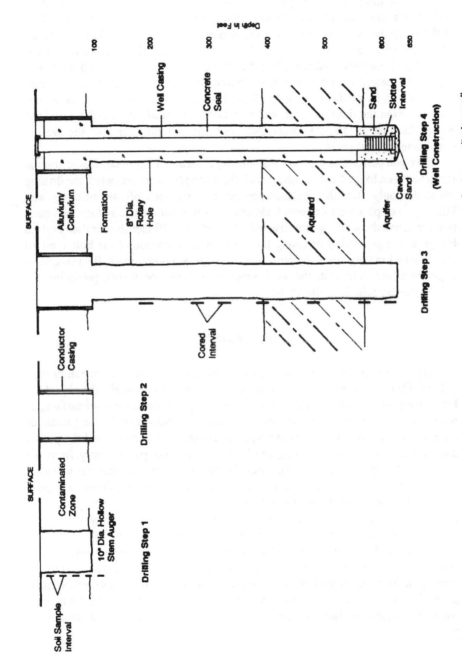

Figure 17   Possible approach to drilling, logging, and installing a deep groundwater monitoring well.

is much sandier and fine-grained than the literature has inferred for this area. Hence, while the borehole is open, you had the foresight to schedule an electric logger to log the hole, so now you will have two complete logs. The logger must perform his task quickly since the circulating mud will erode the aquifer section of the borehole. Well construction must immediately follow the electric logging. When the well casing is lowered you discover that about 10 feet has caved at the base of the hole (now 640 feet). The driller has informed you that while he may be able to clear the lower 10 feet, the borehole may cave to 600 feet (because the aquifer is so sandy) forcing redrilling that interval under flowing sand conditions. Given the time and cost, and now borehole instability, the decision is to accept the loss of 10 feet rather than lose the borehole, and so the well is constructed at that point.

Have you performed the tasks assigned by contract? The contaminated alluvial strata has been logged and chemical analytical samples were collected, the drilling proceeded only after the contaminated zone was sealed off, permeability and lithologic samples were collected, electric logs will fill other lithologic information gaps, and the well was constructed. The well was 10 feet shorter than desired but given the possibility of losing the entire aquifer portion of the hole (and all the time and money spent therein), it seems a good compromise. By using your experience and judgment, the work scope goals have been met, providing the information required for the study.

SUMMARY

Exploratory drilling requires the proper logging of the soil, sediment, or rock collected from samplers or continuous chips by the hydrogeologist. The borehole log is the primary piece of data for making interpretations of the subsurface, as well as locating areas of contaminant occurrence and pathways. Drilling methods have various usefulness in different types of rocks and sediments and the proper use of equipment aids in getting the best data and sample recovery from the borehole. A knowledge of stratigraphy is paramount when working in either alluvial nonconsolidated sediment or lithified rock, since depositional environments are key to understanding lateral sediment changes and predicting sand and clay strata location. Fractured rocks may also have some cyclicity of fracture occurrence or juxtaposition of rocks by faulting, which might be followed borehole to borehole. At times the extrapolation of data from limited boreholes is critical with limited exploration and sampling budgets, and the proper identification of environment from borehole samples will greatly aid in subsurface mapping. This in turn identifies the sand and clay strata and their structure, and suggests possible contaminant pathways that need to be sampled to define the problem.

CHAPTER 4

Groundwater Monitoring Well Design and Installation

INTRODUCTION

Groundwater monitoring wells are data collection points that are installed during the investigation. The wells are the permanent monitoring system and provide the site groundwater potentiometric and quality information used throughout the project's life. Monitoring wells will be used for some indefinite time, so construction materials selection and installation criteria are very important. Consequently, the effort expended in installing the system should be well thought out and constructed carefully. Improperly designed or located wells may lead to numerous problems including improper plume location, cross-connecting aquifers, and questionable flow and quality information. Time and money are wasted on bad data and further cost may be incurred in new wells and sampling costs.

MONITORING WELL DESIGN

Monitoring wells should be located and constructed to collect representative samples of groundwater quality and provide reliable potentiometric data. The design of the well should always reflect the site-specific hydrogeologic environment. Monitoring well design should be initially done in the office, given known or surmised site-specific geologic conditions. Well installation permits may require the consultant to provide a working schematic or conceptual design prior to installation. At times regulatory agencies may review, modify, or even specify monitoring well design as part of a site operating permit, to protect an aquifer given specific groundwater management problems, or as a specific investigation work plan. This predictably creates problems when fine points of design or geologic realities require design changes beyond arbitrary dictums or generalized design in government guidance documents. Finally, the type of contaminants and natural geochemical conditions will have a bearing on materials selection, positioning of screens and casing, and annular seals.

The finalized design will depend on previous experience, regulatory guidance geologic conditions, and professional judgment (Driscoll, 1986). Since all drilling and well completion operations are complex, the on-site hydrogeologist should observe all phases of drilling, construction, and well completion. Cross contamination of aquifers through improper or sloppy construction or in difficult formations could enhance contaminant movement and, without considerable diligence, may happen to the experienced hydrogeologist (Hackett, 1987; U.S. EPA, 1986). Field experience is required and is vital for success in this work. These projects cannot be properly managed from the office alone; they require both experienced field and office personnel to direct the work.

The monitoring well is somewhat similar to a water production well in general design and construction. A borehole is drilled to some depth into the aquifer. The well may fully penetrate the aquifer thickness, but is terminated just into the first underlying aquitard. A casing is lowered into the borehole with screens or slots to allow water entry. Slots are usually openings cut into the casing, and screens are a wire wrap around wire or bars; usually the screen has more open space per linear foot. The space between the borehole and casing is the annular space. The annular space is filled with a sand pack in the area of the slots or screens (except in some hard rock completions, or in a natural sand pack condition), and an impermeable seal is placed atop the sand pack above the screens to the surface to prevent water from entering and moving down the annular space and fouling the well.

Artificial Filter (Sand) Pack Selection

If native aquifer materials cannot be used as a filter pack, an artificial sand pack is usually required. The use of an artificial sand pack is particularly good in the following environments; uniform fine-grained sand; poor cementation of sand; predominantly clay and silt; high degree of stratification in aquifer; and where a long screen interval is used. The sand pack stabilizes the formation (preventing caving) and helps to reduce turbidity while allowing water entry into the well (Aller et al., 1989).

Design criteria for sand packs involve review of the aquifer lithology and proposed well screen size, and guidelines for well design exist. Several methods may be used that were developed for water production wells, but these "standards" must be used with care since a monitoring use is somewhat different (see Driscoll, 1986; U. S. Department of the Interior, 1981; U.S. EPA, 1989). Samples of the aquifer are collected and a grain size sieve analysis of the grain distribution is made. Once the aquifer grain size distribution is known, then the effective size of the aquifer material is estimated and a sand pack is selected. Usually the sand pack must be slightly larger to retain the formation while allowing maximum water entry through the screens of slots. Finally, the sand pack should be clean, typically using 90% quartz to eliminate reactive minerals, and composed of rounded grains that aid increased yield to the well. Aggregate sand suppliers can then prepare the sand to design specifications.

Usually at least one sand sieve analysis should be done for the site and the results used for the monitoring-well sand pack and slots selection. The sand pack design will involve judgment using the sieve data and uniformity coefficient to select the sand filter pack. The uniformity coefficient (Cu) is the ratio of the sieve size through which 60% (by weight) of the material passes to the sieve size that allows 10% of the material to pass. This gives an indication of the uniformity of the formation. If formations are relatively uniform,then a factor between 4 and 10 is selected. Driscoll (1986) recommends that the filter pack for water wells should have a uniformity coefficient of 1 to 3 (multiplying the 70% retained of the finest formation sample by the aforementioned factor; if "uniform" use 4 to 6; if unconsolidated and contains fines then use 6 to 10). This gives a first point on the sand pack curve for grain size, and the sand pack curve may be interpreted (see Figure 1). Other guidance indicates that, depending upon the range of the uniformity coefficient, the sand pack falls into one of three groups which retain 90% of the formation, or slots which do not admit more than 10% of the sand pack (U. S. Department of the Interior, 1981). Thus, following well development, the sand pack should grade from coarser near the screen to finer away from the screen toward the formation.

A dilemma exists when the aquifer strata contains a large amount of fine-grained (silt and clay) sediment dispersed in the formation. In this case, the design criteria could recommend a screen size and sand pack unrealistically small, which inhibits groundwater entry into the well. Consequently, the design criteria and approach must use the sieve sample data and be tempered by professional judgment and previous well design experience. The calculation and data plot discussed above may begin to approach very small sand pack and slots (very fine sand and slots less than 0.010 inches). The design is modified to allow sufficient water entry with a trade-off of some sand and silt pumping, for which the well redevelopment may be needed more often. Once the aquifer conditions and site geology are known, then a similar design may be used for additional wells on that site (providing subsurface conditions are similar). Generally the aquifer does not vary so radically that a completely different well design is needed, although each well must be governed by the conditions in that borehole. When extraction wells are designed, then the well design should be reviewed for that purpose to attain optimum well performance. The design is a general procedure tempered by the professional's experience and past well design experience.

Well Casing and Screen or Slot Materials Selection

Casing materials selection for wells is one of the most important decisions in design for longevity and resistance to site chemistry. Consider that groundwater monitoring wells may be used for periods ranging from months to decades (UST studies and long-term monitoring may last 1–10 years, while RCRA postclosure

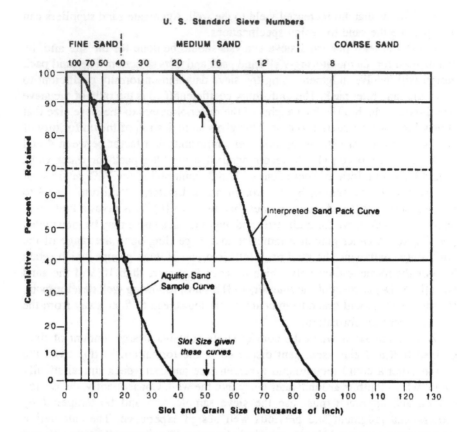

Figure 1 Example of selection of sand pack and slot design. Uniformity coeficient; Cu = D60 ÷ D10, so 0.21 ÷ 0.01 = 2.1; Cu number indicates relatively uniform formation. Factor between 4 and 10 needs to be selected. 4 is chosen because formation sample is interpreted uniform. 70% formation sample grain size retained is 0.015 inches, so 0.015 inches × 4 (factor) = 0.06 sand pack grading (arrow). Based on this information, the well screen or slots should retain 90% of the sand pack so the maximum slot size = 0.05 inches. The design is modified as professional experience and/or formation factors indicate. (Modified from U.S. Department of Interior, 1980, and U.S. EPA, 1989. With permission.)

monitoring could last up to 30 years). This means that the well has to perform properly even though it may be used by several different consultants and even different property owners. Well casing is the pipe that constructs the well, with the screens or slots at the lower end and the solid pipe (sometimes called riser), which is threaded or welded together (epoxy glues are not used because they may desorb similar volatile contaminants found in solvents or fuels).

Numerous articles and tests have been done on various casing materials (see Pearsall and Eckhardt, 1987; U.S. EPA, 1986a, 1992). The materials most used are plastics (polyvinyl chloride [PVC], Teflon®) and metals (steel and stainless steel) according to EPA and a poll by McCray (1986). EPA guidance in the mid-1980s ranked the most used materials in order from "best to worst" as Teflon®, stainless steel, PVC Type 1, low carbon steel, galvanized steel, and carbon steel. McCray's poll revealed that nationwide, consultants responded that PVC was the material of choice by 93%. The most important concept to remember is that each material has some advantages in some environments and possible liabilities in others. Site-specific use is the key.

The materials PVC, steel, and Teflon® each have their use in monitoring at large or small sites, each having its advantages and disadvantages in certain geologic and contaminant environments. Beyond construction material strength and service life considerations, the potential to release or absorb contaminants is the primary design criterion. For example, steel casing used in extremely saline or cathodic environments may release alloy metals and corrode. PVC may soften and decay in environments of high dissolved concentrations or separate phase product. Teflon® may release metals in certain acid environments (Creasey and Dreiss, 1985). Consequently, the casing choice is the one that best adapts to the problems of water quality, geology, and budget (Table 1).

Leaching tests by Sykes et al. (1986) and Barcelona and Helfrich (1986) show that in material exposure experiments, significant differences did not exist for these materials in terms of leaching and adsorption. Thus, the casing selection for some sites with other factors being equal is oftentimes a financial consideration. Since the cost of steel may be four times PVC, and PVC may be ten times that of Teflon®, the project budget factors into the construction criteria. However, the cost basis should be made given the estimated chemical exposure and service life, and to yield the best data.

Length of Well Screens or Slots

The length of the well screen or slot is dependent on the aquifer thickness and contaminant type for site monitoring. The screen should allow the well to yield water with optimal efficiency, sample the contaminants of interest, allow for seasonal groundwater fluctuation, and enable hydraulic head data collection. For example, monitoring wells that are installed to monitor immiscible contaminants should be screened somewhat above the "static" water levels to allow separate phase product entry into the well. Wells may be discretely screened to observe stratified or "sinking" contaminants at different aquifer depth intervals.

Ideally, a monitoring well screen should penetrate the entire aquifer thickness. Usually fully penetrating screens are installed in the initial site investigation where aquifer delineation is done to gather the maximum information for the available budget. Discretely screened wells may be installed for specific depth intervals of interest in later investigation phases. If the aquifer is relatively thin (less than 20 feet thick), fully penetrating wells might be used throughout. This may not be practical in very thick aquifers, and several wells may be needed to cover the thickness. The selection process is a judgment of the site geologist, based on a review of the study data. However, it is always done to maximize data quality and should never provide pathways to aid contaminant movement.

Table 1 Summary of Development Methods for Groundwater Monitoring Wells, (Aller et al., 1989)

Overpumping	Backwashing	Surge block[a]	Bailer	Jetting	Airlift Pumping	Air Surging	Ref.
Works best in clean coarse formations and some consolidated rock; problems of water disposal and bridging.	Breaks up bridging, low cost & simple; preferentially develops.	Can be effective size made for >2" well; preferential development where screen >5'; surge inside screen.		Consolidated and unconsolidated application; opens fractures, develops discrete zones; disadvantage is external water needed.	Replaces air surging; filter air.	Perhaps most widely used; can entrain air in formation so as to reduce permeability, affect water quality; avoid if possible.	Gass (1986)
Effective development requires flow reversal or surges to avoid bridges	Indirectly indicates method applicable; formation water should be used.	Applicable; formation water should be used; in low-yield formation, outside water source can be used if analyzed to evaluate impact.	Applicable.		Air should not be used.	Air should not be used.	U.S. Environmental Protection Agency (1986)
Productive wells; surging by alternating pumping and allowing to equilibrate; hard to create sufficient entrance velocities; often use with airlift.		Productive wells; use care to avoid casing and screen damage.	Productive wells; more common than surge blocks but not as effective.			Effectiveness depends on geometry of device; air must be filtered; crew may be exposed to contaminated water, perturbed Eh in sand gravel not persistent for more than a few weeks.	Barcelona et al.[b] (1983)

				Reference
Applicable; drawback of flow in one direction; smaller wells hard to pump if water level below suction.	Suitable; periodic removal of fines.	Suitable; common with cable-tool; not easily used on other rigs.		Scalf et al. (1981)
		Suitable; use sufficiently heavy bailer; advantage of removing fines; may be custom made for small diameters.	Methods introducing foreign materials should be avoided (i.e., compressed air or water jets).	National Council of the Paper Industry for Air and Stream Improvement (1981)
Development operation must cause flow reversal to avoid bridging; can alternate pump off and on.	Applicable; caution against collapse of intake or plugging screen with clay.			Everett (1980)
Probably most desirable when surged; second series of evacuation/recovery cycles is recommended after resting the well for 24 hours; settlement and loosening of fines occurs after the first development attempt; not as vigorous as backwashing.	Suitable; periodic bailing to remove fines.		Suitable; avoid injecting air into intake; chemical interference; air pipe never inside screen.	
Vigorous surging action may not be desirable due to disturbance of gravel pack.	Method quite effective in loosening fines but may be inadvisable in that filter pack and fluids may be displaced to degree that damages value as a filtering media.	High velocity jets of water generally most effective, discrete zones of development.	Suitable	Keely and Boateng (1987)
		Popular but less desirable; method different from water wells; water displaced by short downward bursts of high pressure injection; important not to jet air or water across screen because fines driven into screen cause irreversible blockage; may substantially displace native fluids.	Air can become entrained behind screen and reduce permeability.	

a Schalla and Landick (1986) report on special 2" valved block.
b For low hydraulic conductivity wells, flush water up annulus prior to sealing; afterwards pump.

Annular Seals

The well annulus between the borehole wall and the casing must be sealed to isolate the screened interval and prevent contaminant entry. Since the seal is required to be "impermeable," the sealing material is usually a portland cement grout, bentonite clay seal, mixture of grout and bentonite, or sandwiched seals. The seal is either poured into the borehole if shallow, or poured through a tremie line or pipe. A tremie line is simply a pipe (or hollowstem augers) through which the seal is poured directly to the bottom of the borehole, filling it from the bottom to top. Since the cement is poured to a point atop the sand pack, it can mix and thin with standing water in the borehole. Hence, sometimes a bentonite clay seal is placed above the sand pack to prevent grout invasion into the sand pack (and into the well screens, a common problem if staff is not attentive to detail during construction). Sometimes a sand spacer is placed from the top of the slots to a point several feet above the last slot or screen if bentonite cannot be used, again to prevent grout invasion. Where approved by the agency, a bentonite gel product may be mixed and pumped as a seal instead of using grout.

Grout is usually a commercial portland cement mix, to which is added bentonite to make it more impermeable. The bentonite–grout mix tends to be 5 to 7% bentonite. Grout mixtures may be specified by the regulating agency, or can be dictated by site-specific conditions (such as reactive minerals that do not allow a standard cement to set). Grout is prepared by aggregate or cement plants and usually the required mix can be purchased in bags and mixed on site. If exceedingly long seals are required, then a mixer truck may deliver directly to the site. Most grout seals settle a few feet after pouring and may need to be topped off prior to setting the box, and grouts typically set up in 12 to 24 hours, reaching their strength in 21 to 28 days.

Surface Completion

The well surface completion provides ground surface access and security at the well head. Most monitoring wells are completed below the ground surface and accessed through a traffic-rated concrete box set in the pavement. If well casing is completed above ground, then marking pipes and traffic barriers must protect the casing. The surface portion of the casing is contained in a "stovepipe" or steel pipe sticking out of the ground within the marking protection pipes. As with a surface completion, the well head is accessed through a locked cover to prevent unauthorized entry and vandalism. A metal tag containing the well construction information may be attached to the casing inside the box. Final completions may be further tailored to the needs of the site, client, or regulatory agency requests.

Summary

Monitoring well design must take into account site geology at the well location, type and position of contaminant, input from regulatory agencies, and previous well installation experience. Contingencies for well design field modification should be anticipated prior to going on site, and if subsurface conditions are significantly different from those expected. The screen length of each well should be site-specifically determined according to the aquifer conditions observed and contaminant type. The well sand pack should be designed for the aquifer texture and may be used elsewhere on site, providing subsurface conditions are similar. An impermeable seal is placed atop the sand pack poured into place by tremie line, and finished at the surface with a secure locking box. The final well design and care of construction takes these considerations into account so the monitoring well is usable for the service life.

EXAMPLE OF MONITORING WELL INSTALLATION INTO THE "A" AQUIFER

The following example discusses the approach for installation and basic construction sequence of a well into the uppermost water-bearing strata, or "A" aquifer at a site (see Figures 2, 3, and 4). We will assume that the borehole had been advanced to the desired depth, lithologic logging has determined the screened interval depth, and contaminant type and geology were factored into the casing materials selection. Finally, we assume that the hydrogeologist, has finalized the design and the proper material is cut to his specifications. The field geologist, hydrogeologist, or engineer at the drilling site must take ultimate responsibility for the proper well construction completion.

The site hydrogeologist must view the casing prior to installation to look for potential damage and contaminants, and clean it if needed. Since casing is commonly manufactured and sold in 5-, 10-, or 20-foot lengths, the casing is threaded together and lowered into the borehole in those increments. As the casing is threaded together, the threads should be checked and secured to the end of the thread, lest an "airfall" of the casing arise from the casing parting and falling into the borehole. (This can necessitate a "fishing trip" to remove the damaged casing, which sometimes cannot be retrieved, causing the loss of the borehole and redrilling). The end of the slotted or screened casing is capped with a plug or endcap so water does not pump sediment at the end of the casing. Tension is kept on the casing string to prevent kinking, and well centralizers may be used to keep the casing centered in the hole. Once all the casing has been lowered into the borehole, then the annular space is filled with sand pack and the seal.

The sand pack installation must be done carefully to avoid sand bridging and ensure the sand is properly delivered around the screen. The sand flows down by a tremie pipe (often the hollowstem auger) or may be sent by airfall down the borehole in shallow (less that 40 feet. The sand pack will fill the annulus, which usually is a minimum of 2 to 4 inches between the casing and borehole wall. Once emplaced about the screened interval, the sand is typically filled to a point about 2 feet above the top of the screens or slots. This ensures the screen/slot interval is completely covered and creates a spacer above the last screen opening and the seal.

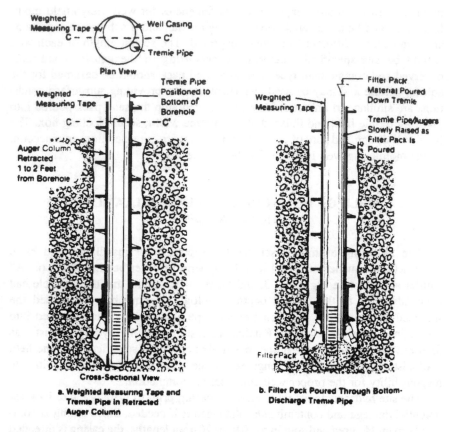

Figure 2   Installing well filter sand pack using hollowstem augers as tremie line. (From Aller et al., 1989. With permission.)

The annular seal will isolate the well screen in the aquifer and preclude communication with other aquifers. A bentonite seal (about 2 feet thick) may be placed above the sand pack prior to grouting the annulus. This is very important to prevent grout invasion of the sand pack, which would render the well useless. Using bentonite at this point also forms an annular seal and isolates the aquifer interval while the borehole is still open. The bentonite product may be pellets, chips, or powder that hydrates and expands on contact with water. If bentonite cannot be used due to extreme sand pack depth (perhaps over 150 feet), then a 5- to 10-foot sand spacer may be placed above the sand pack to prevent grout invasion.

Once the bentonite or sand spacer is placed, a grout seal is tremied into the annulus and seals the aquifer. The grout is pumped downhole until the space is filled to the ground surface. Grout setting (heat and weight) may adversely affect nonsteel casing, and the grout may need to be placed in intervals or lifts to minimize problems. Again, the care in construction should be the same as that of any other part of the geologic and hydrogeologic study. The site hydrogeologist should observe the entire construction sequence and note additional material

Figure 3A Casing lowered into hollowstem auger. (Photo by the author.)

Figure 3B Pouring sand pack into annulus using hollowstem as tremie line. (Photo by the author.)

Figure 3C   Pouring bentonite pellets to begin annular seal above sand pack. (Photo by the author.)

needs, borehole instability, and other construction problems so they can be avoided at other holes. This is also very important if other parties or the regulating agency raise liability questions regarding proper construction.

Finally, the surface completion must be done to protect the well surface casing from flooding and vandalism. A secure cap with locking device should cover the wellhead, which in turn is contained in a traffic-rated vault box if completed at grade. If the wellhead is completed above ground, a concrete pad surrounded by guard posts should be used whether in a field or parking area, since jarring the casing can snap it off near the surface. The well access should be graded slightly so positive drainage is provided away from the well. Additionally, well completion information, such as depth, screened interval, date of completion, and well permit number, may be stamped on a well tag attached to the casing in the box (see Figure 4, "A" detail).

MONITORING WELL DEVELOPMENT

All wells must be developed upon completion. Well development accomplishes the following: (1) it clears suspended particles from the water column; (2) it removes mud cake, crushed rock, and particles from drilling action and smeared materials on the borehole wall, particularly in mud rotary drilled boreholes; (3) it grades the sand pack from coarser near the screen to finer toward the formation, making more complete contact providing the best hydraulic connection between the casing and aquifer. Well development should be done to provide the best result without damaging the well casing or creating voids, or blocking porosity in the aquifer (see Table 1; Figure 5).

The most common development processes are surging, bailing, and jetting (see Driscoll, 1986; Keely and Boateng, 1987). The surge method uses a block that fits snugly into the casing and is raised and lowered in the screened interval (see Figure 6). Surging causes water to move in and out of the screens or slots

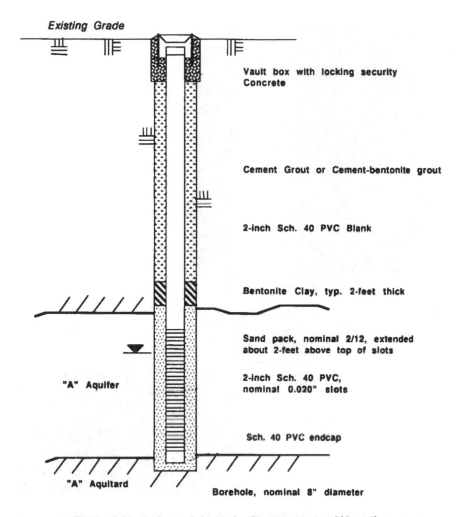

Existing Grade

Vault box with locking security
Concrete

Cement Grout or Cement-bentonite grout

2-inch Sch. 40 PVC Blank

Bentonite Clay, typ. 2-feet thick

Sand pack, nominal 2/12, extended
about 2-feet above top of slots

2-inch Sch. 40 PVC,
nominal 0.020" slots

Sch. 40 PVC endcap

"A" Aquifer

"A" Aquitard

Borehole, nominal 8" diameter

Figure 4 Monitoring well design for the uppermost or "A" aquifer.

and entrains clay and silt with fine sand in the water. The surge block is removed and a bailer is lowered and bails out the turbid water. The surge block is lowered and the process is repeated and continues until the water becomes clear and the pumping of fine sand is minimal. The process is usually vigorous, but should not be to the point of causing formation damage.

Bailing involves lowering a bailer into the borehole and lifting water to the surface causing the fine sand, silt, and clay to become suspended and then removed from the casing. Unlike surging, this process is "one-way" in that the water moves only from the aquifer into the casing. Surging is more active and helps to remove fines from the sand pack and slightly in the aquifer. However, this process could be harmful if the aquifer is fine-grained or contains much disseminated clay. The author's field experience with fine-grained aquifers (containing abundant and disseminated silt and clay) is that water clarity is achieved

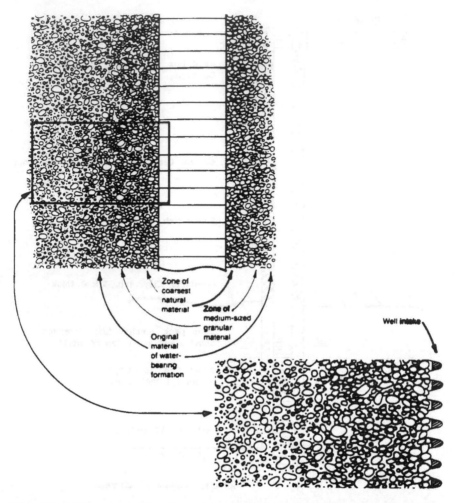

Figure 5 Well development and filter pack gradation in a naturally developed well. A similar gradation is the goal in a filter pack well, with gradation coarsening in the formation outside toward the filter pack. (From Aller et al., 1989. With permission.)

with difficulty, if at all. Gently bailing this type aquifer to initially remove drilling muck and set the sand pack develops the well and development is stopped when fine sand pumping is minimal. However, in some cases, water may be slightly turbid because of that formation.

Jetting involves lowering a device that sprays or jets water into the well at high speed, clearing the drilling muck and cleaning the screens. The jet may be used with a surge block, and following jetting, a bailer removes the entrained material. Jetting can be effective, but may cause some damage by shooting particles into the screens, plugging particles or air in the porosity (Keely and Boateng, 1987).

When a well is properly developed, aquifer water should enter the screens and over the well's use, and well siltation should be minimal. Depending on the

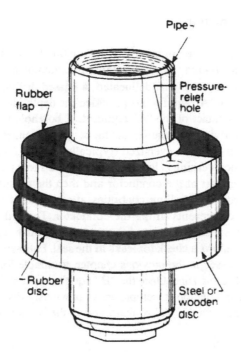

Figure 6 Surge block developing tool. (From Driscoll, 1986. With permission.)

type of formation being drilled, the ultimate degree of well development may be based on professional experience. Turbid water at the end of well development may be unavoidable given a fine-grained lithology. Notes on well development and a log of the process should be kept in the field. This data can be very useful over the monitoring well use and may help answer questions of well yield and data peculiar to one portion of the formation or site. Over time, the well may need to be periodically redeveloped. This would involve the processes described above, to remove the accumulated sand and silt pumped over the well's life. One indication of the frequency of redevelopment is to check sampling forms for depth of well casing soundings. If the well bottom becomes "shallower" and screen interval decreases (say 20% of the interval), then the well should be redeveloped. As with other aspects of this work, the need is often site-specific and well-specific.

WELL COMPLETIONS FOR DIFFERENT MONITORING SITUATIONS

The generalized well installation discussed above may be used to monitor the uppermost aquifer (or first encountered "A" aquifer). However, monitoring wells have to be placed in deeper aquifers to search for different potentiometric surfaces and in a thick aquifer for stratified contaminant flow. Several different completion techniques have been worked out to provide this coverage and include the next underlying or "B" aquifer, nested completion, and well clusters.

"B" Aquifer Completion

The "B" aquifer completion involves constructing the well to the next under-lying water-bearing stratum, through a known contaminated stratum. Since the upper aquifer has been somewhat delineated, a phased drilling approach is used. The delineation of the "A" aquifer has provided the depth interval of "A" and the depth to the upper contact of the "A" aquitard. A borehole is advanced through the "A" aquifer verifying the aquifer contacts, and advanced about 1 foot into the aquitard. This borehole is oversized (typically 10- to 12-inch diameter), into which a steel conductor casing is lowered and pressed 1 to 2 feet into the aquitard. This seals the lower part of the conductor and then the outside annular space is cemented (see Figure 7). Drilling continues after about 24 hours to allow the cement to set, which prevents lifting the conductor. The drilling muck and any accumulated water is then bailed out of the conductor, which removes the contam-inated material and allows a check for leaks in the seal. Following this, the borehole is advanced as discussed in the previous chapter, to sample for contaminants and identify the water-bearing zone. Once the "B" aquifer upper and lower contacts are defined, then well installation is completed as previously discussed. The annular seal is completed through the "A" aquitard and in the conductor casing.

Nested Completion

A nested completion usually involves two or more monitoring wells com-pleted in the same borehole. This may be an efficient way to complete wells that are installed to observe stratified effects in the aquifer (see Figure 8). However, these can be complicated and difficult construction problems, limited by the available space in the single borehole and adequately setting the seals between the screened intervals of the individual wells. The seal problem is especially important because if groundwater circulates in the borehole, the purpose of the nest is defeated. Other difficulties may arise if caving formation conditions exist, or erosion is caused by excessive mud circulation, or casing trips occur. These completions may be somewhat easier for piezometers used only for potentiometric measurements in one aquifer, but can be headaches for monitoring, so well clusters might be used.

Well Clusters

Well clusters are individual wells constructed in close proximity to each other (within 5 to 15 lineal feet). The wells are constructed in steps as in the "A" aquifer case, but well screens penetrate different depth intervals. Thus, the wells monitor discrete depths within a single aquifer to observe contaminants (see Figure 8). While clusters are typically installed in groups of three, four, or five, an initial borehole may be drilled from which the stratigraphic and sampling data are collected. A significant lateral difference in site stratigraphy is not likely over a very short distance, so the wells are installed by drilling directly to the target

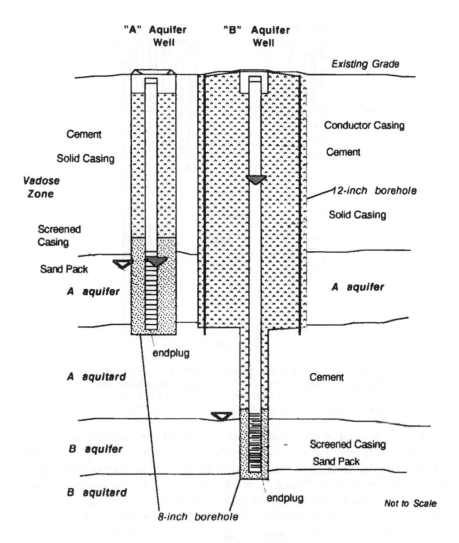

Figure 7 "B" monitoring well completion.

depth and completed. Of course, contacts are verified, or limited sampling done to assure the desired well location, or altered as site needs require. Data collected from clusters is typically very good and eliminates the construction questions of nests.

WELL ABANDONMENT

At some point, monitoring wells may be abandoned (properly destroyed) when they are no longer needed, have construction problems, or are part of a site closure. It is wise to abandon wells when they are no longer needed to avoid

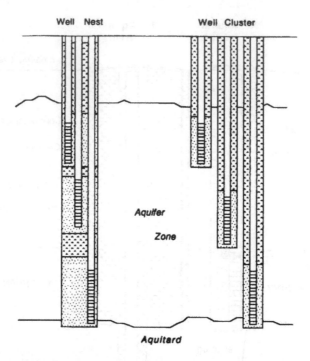

Figure 8 Well nests and clusters. Well nests are individual monitoring wells constructed in the same borehole. Well clusters are individual wells constructed in individual boreholes in close proximity to each other. These well arrays are used to observe specific aquifer intervals and stratified contaminant flow.

inadvertent contamination, vandalism (including adding contaminants), owner liability, and "losing" the well. Sites are often redeveloped or change ownership, and the recorded location of the well may not be known; depending on agency records, the site may be plowed or paved over through inaccuracy of site maps.

Wells are usually abandoned by drilling them out, or filling the interior of the casing with a seal. The ultimate type of abandonment procedure depends on the well condition and depth, agency requirements, and location. A well may be destroyed by drilling out the well, casing, sand pack, and seal, then filling the resulting borehole with cement or other seal. This means that the construction information of the well should be known prior to abandoning so the materials and depth are known. The well may be sampled one last time, and the depth to water and total depth should be measured and recorded. If the well is old or casing failed, then the discrepancy of depths may be resolved with the original well log, detail, and permit. If the well is steel-cased, drilling is not viable, and the well is filled.

A cement slurry then fills the well casing from the bottom up. A tremie line may be used or a seal may be pressure grouted. If possible the casing is pulled from the ground. If pulling cannot be done, the casing may be punched full of holes or ripped with a casing knife to allow the grout to invade the sand pack. The borehole should be filled so that it does not become a conduit for fluid movement. A check on the amount of grout to use is to calculate the volume prior to starting so the proper amount of material is used. If a bridge forms, then the process must stop and the bridge must be removed so the seal it properly and completely placed. Upon completion of the destruction, a permit or other document of the abandonment should be prepared and filed with the supervising agency. This provides a permanent record of the destruction.

CHAPTER 5

Groundwater Monitoring Well Sampling

INTRODUCTION

Groundwater monitoring well sampling is one of the most important tasks performed during a contamination investigation. Together, the exploratory borehole log and the water quality information provide the two basic sources of observed subsurface data. Monitoring well sampling provides the groundwater geochemical and contaminant chemistry information for the problem under consideration. Many times, the investigator will be dealing with contaminants at the parts per million (ppm, milligrams per liter, mg/l) or part per billion (ppb, micrograms per liter, μg/l) concentration level. Consequently, the possibilities of errors in data collection are enlarged and care and quality control procedures must be used when obtaining samples.

The consultant is often asked to ascertain whether or not a contamination problem is present, and the chemical data will dictate the presence, nature, and type of problem. When the information is reviewed by the regulating agency, the presence of one type of contaminant, even at very low concentrations, may initiate a costly investigation, compliance, and remediation. Hence, after properly constructing the monitoring well, the water sample collection procedures must be carefully thought out and executed to obtain reliable data. Finally, site specificity cannot be overemphasized. The aquifer chemistry may be very subtle or can change seasonally or throughout project life, and these variations should be observed over time. At times, data from only several sampling events may be present, and judgments may need to be made on four or eight quarters of data to establish trends. Hence, the sampling procedures should accurately define aquifer hydrogeochemistry.

SAMPLING PLANS AND PROTOCOL

Usually a sampling plan and procedures, commonly called the sampling protocol, are prepared prior to field investigation. A general sampling protocol,

Table 1   Generalized Groundwater Sampling Protocol Procedures (EPA 1987)

Step	Goal	Recommendations
Hydrologic measurements	Establish nonpumping water level.	Measure the water level to ±0.3 cm (±0.01 ft).
Well purging	Remove or isolate stagnant H$_2$O which would otherwise bias representative sample.	Pump water until well purging parameters (e.g., pH, T, Q^{-1}, Eh) stabilize to ±10% over at least two successive well volumes pumped.
Sample collection	Collect samples at land surface or in well-bore with minimal disturbance of sample chemistry.	Pumping rates should be limited to ~100 mL/min for volatile organics and gas-sensitive parameters.
Filtration/preservation	Filtration permits determination of soluble constituents and is a form of preservation. It should be done in the field as soon as possible after collection.	Filter: Trace metals, inorganic anions/cations, alkalinity. Do not filter: TOC, TOX, volatile organic compound samples; other organic compound samples only when required.
Field determinations	Field analyses of samples will effectively avoid bias in determining parameters/ constituents which do not store well; e.g., gases, alkalinity, pH.	Samples for determining gases, alkalinity and pH should be analyzed in the field if at all possible.
Field blanks/standards	These blanks and standards will permit the correction of analytical results for changes which may occur after sample collection: preservation, storage, and transport.	At least one blank and one standard for each sensitive parameter should be made up in the field on each day of sampling. Spiked samples are also recommended for good QA/QC.
Sample storage/transport	Refrigerate and protect samples to minimize their chemical alteration prior to analysis.	Observe maximum sample holding or storage periods recommended by the Agency. Documentation of actual holding periods should be carefully performed.

using suggested procedures based on government guidance documents, is used by consulting firms as their basic internal procedure document. Other sampling plans are written specifically for projects or specific contaminant types developed from the basic document. All plans are usually derived from the government guidance, and amended by local agencies (see Table 1, Figure 1). At times the investigation and sampling plans may be reviewed by regulators and other involved parties, and may take months and even years before a procedure consensus is reached. It is important to have sampling procedures that are similar to industry or regulated standards and provide the needed documentation for collection and handling from the field to the analytical laboratory. Regardless of project size or complexity, the level of care should be the same when collecting any subsurface data because it could be legally challenged at some future time.

Numerous agency-prepared groundwater sampling guidance documents are available and used to design and write sampling plans (for example Nielson and Johnson, 1990; State of California, 1986; U.S. EPA, 1985, 1986, 1987), and the

Type of constituent	Example of constituent	Positive displacement bladder pumps	Thief, in situ or dual check valve bailers	Mechanical positive displacement pumps	Gas-drive devices	Suction mechanisms
		INCREASING RELIABILITY OF SAMPLING MECHANISMS →				
Volatile Organic Compounds Organometallics	Chloroform TOX CH_3Hg	Superior performance for most applications	May be adequate if well purging is assured	May be adequate if design and operation are controlled	Not recommended	Not recommended
Dissolved Gases	O_2, CO_2	Superior performance for most applications	May be adequate if well purging is assured	May be adequate if design and operation are controlled	Not recommended	Not recommended
Well-purging Parameters	pH, Q^{-1} Eh	Superior performance for most applications	May be adequate if well purging is assured	Adequate	May be adequate	May be adequate if materials are appropriate
Trace Inorganic Metal Species Reduced Species	Fe, Cu NO_2^-, S^-	Superior performance for most applications	May be adequate if well purging is assured	Adequate	Adequate	Adequate
Major Cations & Anions	Na^+, K^+, Ca^{++} Mg^{++} Cl^-, SO_4^-	Superior performance for most applications	Adequate May be adequate if well purging is assured	Adequate	Adequate	Adequate

← INCREASING SAMPLE SENSITIVITY

Figure 1   Matrix of sensitive chemical constituents and various sampling devices (From U.S. EPA, 1991. With permission.)

necessary components of the sampling protocol are well established. The basic plan should define procedures employed to collect all the related information, with the data logged and managed through paperwork, including chain-of-custody and data-collection forms. The same data-collection techniques and the proper sampling handling will follow the data through the life of the project. The information will ultimately create a project database to monitor changes in groundwater elevation, flow direction, and chemical evolution. This data would also be used to ascertain the contaminant extent and for design in site remediation procedures.

SELECTION OF CHEMICAL ANALYTICAL LABORATORY

The selection procedure of the chemical analytical laboratory should be as rigorous as the sampling plan preparation. The site data will be a function of the chemical analysis as well as sample collection, so the laboratory must be able to perform the required analyses according to approved analytical techniques. In some states, analytical laboratories are state-certified for types of analytical procedures, and local regulatory agencies may require that certification be a portion of the sampling plan. The use of EPA methods of chemical analysis are almost always required for environmental studies (such as for fuels, solvents, pesticides, or trace elements). Also, laboratories may have a specialty analyses set, as for example pesticides. These may or may not have "approved EPA methods" and the analysis method may have been developed by the manufacturer of the product. If that is the case, that method could take the place of an EPA method.

At times the laboratory nearest the site may not perform the chemical analysis, and the samples must be sent to another town or another state. Contingencies for rapid overnight shipment and handling must be a part of laboratory service, if needed, since samples have a limited storage time prior to actual analysis. The laboratory must be able to document in-house quality control and assurance for sample handling and analysis, and provide the record-keeping and raw data files to check the veracity of samples when data reliability is questioned. Finally, the laboratory chosen should be used for the duration of the project, so differences in quality control resulting from changing laboratories does not become a problem (if needed, split samples can be used to check accuracy).

Quality Assurance and Quality Control Objectives

Quality assurance and quality control (QA/QC) procedures are established to obtain field data for water quality in an accurate, precise, and complete manner so that information is accurate and representative of actual field conditions. This involves both field and laboratory procedures for internal and external checks on data collection. The data collected are usually checked for accuracy, precision, completeness, comparability, and representativeness (similar for analytical laboratory procedures). These terms are discussed in guidance documents and have been extensively presented in government publications, including some codes and regulations of various federal, state, and local agencies (see for example U.S. EPA, 1985a). Definitions for these terms as applied to groundwater quality sampling follow:

Accuracy —— the degree of agreement of a measurement with an accepted, referenced or true value;

Precision — a measure of agreement among individual measurements under similar conditions, usually expressed in terms of standard deviation;

Completeness — the amount of valid data obtained from a measurement system compared to the amount expected for project data goals;

Comparability — expresses the confidence with which one data set can be compared to another;

Representativeness — a sample or group of samples that reflects the characteristics of the media or water quality at the sampling point and how well the sampling point represents the actual parameter variations under consideration.

These criteria are similar to laboratory QA/QC and can include the analyst technique, cleanup procedures, instrument calibration, equipment detection limits, and instrument maintenance. Since the sampler and laboratory will interact closely, the same QA/QC procedures are used to collect the best groundwater data. The hydrogeologist may wish to review the laboratory certifications, tour the lab, and discuss the crucial aspects of the project to be sure the chemists and support people can maximize the data quality for the job.

MONITORING WELL SAMPLING — PREFIELD SAMPLER PREPARATION

The sampler must prepare for a sampling session by decontaminating and calibrating his equipment. Today, sampling equipment is built of "inert" materials such as stainless steel and Teflon® to minimize potential contaminant problems from equipment parts. Many equipment manufacturers and types of water sampling are available today (Parker, 1994; refer to the appropriate industrial suppliers). Equipment cleaning is accomplished by decontamination procedures, such as washing the bailers, pumps, hoses, lines, and related gear with trisodium phosphate or alconox-type soaps, followed by a distilled or deionized water rinse, and occasionally a solvent rinse, as the protocol directs. Steam cleaning of equipment may be used as required. If cleaning and decontaminating cannot satisfactorily clean the equipment, then it should be replaced. Equipment used to measure depth to groundwater, separate product phases, bailers, and pumps should be clean and protected from inadvertent contamination. Water quality purge parameters of pH, conductivity, dissolved oxygen meters, and so on should be checked daily for calibration accuracy according to the individual manufacturer specifications. Since these meters yield basic general information on ambient aquifer conditions for each sampling event, proper calibration of field gear is vital.

SAMPLE BIAS

Sample bias problems may happen at any time and continuing checks on sampler technique, equipment, and chemical analytical procedures is an ongoing task. Sample bias may be difficult to determine given the large potential for something to go wrong. Individual sampling technician training and techniques may vary widely among individuals or companies. This may include the local and state agency oversight in which the consultant practices, which may have special local hydrogeologic conditions. Different sampling methods and measuring devices have their individual advantages and disadvantages. The methods should be appropriate for contaminant type, whether volatile organic chemical, trace elements, pesticides, etc., for minimum headspace, proper container preservation if needed, and so on. The site-specific sampling method protocol must be carefully followed each time so that possible sampling bias will not affect the reproducibility of the results. In other words, one does not want field sampling error to affect laboratory results (see Table 1; Figures 1 and 2).

Sampling bias could result from a number of possible sources, which include but are not limited to the following: improper purging of the well; sampling wells in low permeability aquifers; improper cleaning of equipment, improperly prepared or contaminated containers, aerosol sources, and sampling from contaminated wells prior to clean wells. The sampling protocol should try to take these potential problems into account so they can be avoided. For example, if oily contaminants are present, steel sampling gear may be an option because plastics and Teflon® may be difficult to clean. If sampling equipment cannot be cleaned, then it should be replaced with new or acceptably decontaminated equipment. At times there may be aquifer-specific problems (such as low yield or turbidity). Usually most site-specific problems can be worked out and the overall protocols modified for those site wells following the first sampling rounds.

Checks on possible laboratory error or bias use sample blanks. The blank checks on both the laboratory and sampler QA/QC for trace amounts of contaminant and possible lab error. Thus, the blanks are very useful to check on overall data accuracy. Usually, a certain percentage of blanks are factored into the analytical program; this is determined on a rate-specific basis or may generally range from 5 to 10% of the total number of samples. Additional random blanks may be run from time to time on the sampler for equipment cleanliness, and spiked blanks (blanks containing a known quantity of contaminant) can be sent to another laboratory to check on the accuracy of the actual analysis. The most common types of sample blanks are:

Trip Blanks are used for purgeable organic compounds only, and are sent to the project site and travel with the collected samples. Trip blanks are not opened and are returned and analyzed with the project samples as a check on the laboratory.
Field Blanks are prepared in the field with organic free water. These samples accompany the project samples to the laboratory and are analyzed for specific chemical parameters unique to the site where they were prepared.

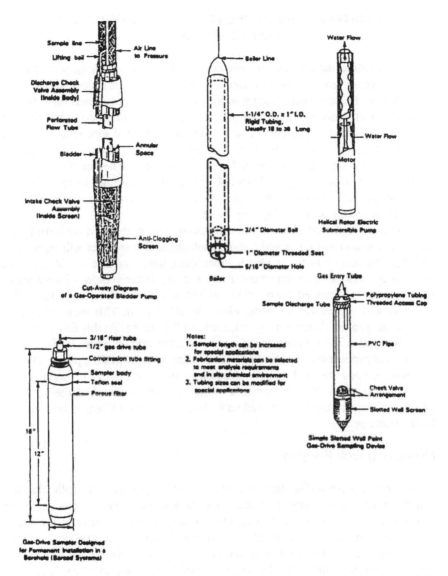

Figure 2 Common groundwater sampling devices. (Modified from U.S. EPA, 1987c. With permission.)

Duplicates are collected as "second samples" from a selected well or project site. They are collected as either split samples (collected from the same bailer volume or pumping discharge), or as second-run samples (separate bailer volumes or different pumping discharges) from the same well. A portion of the duplicate can be analyzed at different labs to check overall accuracy.

Equipment Blanks are collected from the field equipment rinsate as a check for decontamination thoroughness of the sampler.

EXAMPLE — FIELD PROTOCOL FOR SAMPLING A
MONITORING WELL

The sampling data forms are prepared and the well sampling plan usually samples from the least to most contaminated well. Again, this is done to minimize the potential of cross-contaminating wells, and should improve each time the site is sampled and familiarity increases with the site. If very long-term sampling for required monitoring, a prudent course is to install dedicated sampling devices in each well to lessen potential well cross contamination so that equipment decontamination and well disturbance is further minimized. Whether the system is composed of bladder pumps, submersible pumps or some other system, the cost of installation and maintenance is typically offset by reducing suspect data that may initiate redundant sampling (see Table 1).

The sampler should arrive at the monitoring well with clean equipment and properly prepared and preserved sample containers. The sampler will begin data collection by recording the project number, date, time, and site conditions such as weather, well security, and vandalism, and any other required information. Once the well is uncapped, the depth to water is measured to the nearest hundredth of a foot and logged for the water elevation calculation. This measurement is taken at the project datum survey tick mark on the casing. At this time, the well may also be checked for presence of a separate contaminant phase. The apparent thickness of product may be checked with a bailer or water-finding pastes. Optical electronic probes are used that simultaneously measure depth to "product" and then depth to water. Where there is floating product, a check on the apparent product thickness may be measured with clear disposable bailers to measure the bailed thickness.

Monitoring Well Purging

Following these initial steps, the well is then purged prior to collecting the actual sample. This is very important since the geochemistry of the water in the well casing may change due to stagnant conditions and exposure to the atmosphere. The sampler should calculate the required volume that must be removed to draw aquifer water into the well (see Figures 2 and 3). Hence, the well casing depth and diameter and borehole diameter should be known (to take into account the sand pack). Most regulating agencies recognize three to four borehole volumes as a sufficient quantity for removal, although additional volumes may be required by local agencies or project specific needs. However, there are no set criteria for using a set number of purge volumes, and specified purge volume numbers are misleading and could yield questionable data (U.S. EPA, 1985a). The purge volume is specific to each well and its aquifer hydraulic conditions.

Schmidt (1982) reports that experience with monitoring wells in the southwestern United States showed 30 to 60 minutes of pumping at rates of 20 to 50 gallons per minute are needed before chemical parameters stabilize in highly transmissive alluvial aquifers. This may represent an extreme case for purging. In low-permeability aquifers, large purge volumes are impractical; one purge

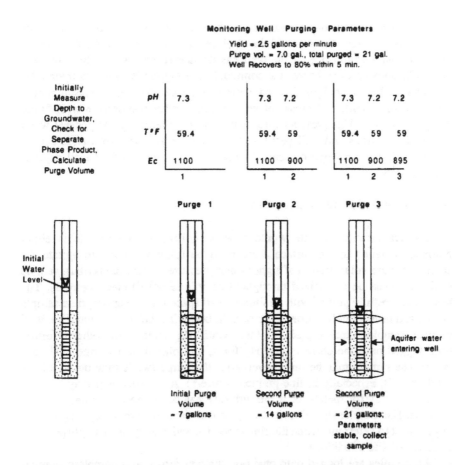

Figure 3 Monitoring well sampling. The well is sampled when the depth to water measurements have been made and the purge volume calculated. The values of parameters stabilize over two consecutive purges and when the well recovers to 80% of the initial water level, the sample is collected. This may happen in three purge events; however, purging is well-specific, as is aquifer recovery.

volume may exceed recharge, and the well must recharge over a long time. Hence, relatively low purge rates are recommended to maintain stable pH and electric conductivities. Wells should be purged according to the site-specific hydraulic well performance for representative sample collection (Gibs and Imbrigiotta, 1990). Unless minimal purge volumes are removed, only casing storage or "stagnant" water will be sampled, which is not representative of the aquifer (Driscoll, 1986).

While the purge volumes are being removed, the basic physical parameters of pH, conductivity, temperature, and dissolved oxygen are monitored to help judge when the aquifer water is entering the well (see U.S. EPA, 1985a). Usually these parameters fluctuate during purging, finally settling to steady value or range of values. For example, typical parameter stabilization ranges may be tenths of degrees temperature Fahrenheit, 0.1 pH unit, and 10s or 100s of conductivity units. Once relatively constant values are observed, the sampler may assume the aquifer water enters the well. If the well recharges quickly, then the sample collection may commence. If the well has dewatered somewhat, then the well should be allowed to recharge, and commonly the sample is collected upon 80% recovery to the originally measured static water level. This information should also be used to build a database of well parameters for comparison with future sampling rounds. This provides information on long-term aquifer general water quality fluctuations and well performance and general long-term trends can be established, such as hydrograph, fluctuations in yield, and overall aquifer geochemistry.

Groundwater Sample Collection

The groundwater sample would now be collected. The sample is either pumped or bailed to the surface. Care must be taken so as not to agitate the sample to cause volatilization of higher vapor pressure contaminants (thus, sample cavitation must be minimized throughout sample retrieval) (see Figure 1). The sample is carefully poured with minimum head space into the appropriate sample bottle, and immediately capped. The sample bottle is labeled with the date, well number, sampler name, project or other identifying number, and other information, logged onto the chain-of-custody form, and placed in a refrigerated container. The sample may be turbid even after the purge (wells may pump a little sand and silt, especially in fine-grained aquifers). A filtering step may be used when analyzing for metals, or if the turbidity has been known to interfere with analysis. The sampler should review the protocol and data with the hydrogeologist or geologist to determine when filtering is needed and to adapt the sample protocol to observed site conditions.

All samples are logged onto chain-of-custody forms so a complete sample-handling record is kept to ensure proper transport and care. Each person responsible for the samples must sign, with date and time, for sample custody. Since proper handling and procedures are vitally important, this becomes a legal document so all samples can be tracked through the project life for regulatory needs and any legal challenges to sample data. Additional information regarding sample analysis method and detection limit, required turnaround time, and so on may be entered on the form.

Once complete, the sampler decontaminates his equipment, recaps and locks the well head, decontaminates the equipment, seals the purge water drum, and moves to the next well.

Sampling "Low Permeability" or Slowly Recharging Aquifers

Sampling wells in low-permeability environments may cause peculiar problems since these water-bearing strata are not usually aquifers in the classical sense. Typically, these aquifers yield low quantities of water, and the fine-grained material (silt or clay) causes the well water to be turbid. When the sampler performs the well purge, the well may dewater and not recharge for minutes, hours, or even days. Similar problems may arise when sampling from aquifers during periods of drought. Sample protocol must be adapted for the low recovery of the monitoring wells and samples collected as the hydrogeology dictates, and may require departure from generally accepted sampling procedures. It may even call into question the definition of "monitoribility" and require interpretation of long duration flow situations given long sample retrieval intervals under regulated sites (Marbury and Brazie, 1988). Hence, the well performance and geology create problems for traditional concepts of monitoring set forth by regulation. Needless to say, negotiation with interested parties and regulating agencies will be required for a consensus on sampling methodology and schedules for these case.

Possible solutions to these problems may have the sampler retain the purge water if the well yield is extremely low, and under special circumstances use this as the sample analysis. If the well recharges over a 24-hour period, then the sampler may return and collect that recharge as the sample. Several samples may need to be collected and a range of chemical quality used for the site background or to mark contaminant presence. It is possible that the sample will be turbid given these formation conditions, and field sample filtering may be necessary. Finally, if the well does not recharge, sampling is canceled for that time interval, and the sampler must wait for the aquifer recharge and the next scheduled sample collection date.

The slowly recharging aquifer can present problems in the accuracy of aquifer chemical conditions. Volatilization can occur from water trickling through the sandpack and casing. McAlony and Barker (1987) report that volatile compound losses of 10% may result in 5 minutes due to recharge water trickling through dewatered sandpack. It may not be possible to collect "representative" samples in the currently understood sense, and the aquifer chemistry may not be comprehended even after numerous sampling rounds.

These problems have to be mediated among the client, consultant, and regulator for agreement on the sampling protocol and criteria for a specific site. Since the site hydrogeology ultimately dictates the nature of the groundwater occurrence and chemistry, flexible protocols may be needed. In this way, the maximum amount of data can be collected to study a site at which groundwater information is difficult to acquire. Interpretation of chemical analyses from samples collected under these conditions should be carefully reviewed. The data accuracy confidence may be low, especially when ascertaining the presence or absence of contaminants that were thought to occur at very low concentrations.

LIQUID SAMPLING IN VADOSE (NONSATURATED) ENVIRONMENTS

At times the need for collection of liquid samples from nonsaturated environments is needed in landfill or land treatment sites to track possible contaminant

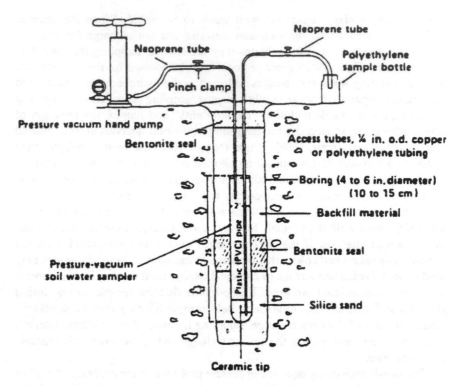

Figure 4   Pressure vacuum lysimeter for unsaturated sampling. (From Wilson, C. G., 1980. With permission.)

movement through the vadose zone toward the aquifer (U.S. EPA, 1985). The technology for this type of sample collection has been available for years and was developed by agricultural science to study water movement in crop root zones. The sampler is the pressure-vacuum lysimeter, which operates on the principle of using soil suction to cause movement of liquid adhered to soil particles into the sampler (see Figure 4). Complete reviews of vadose sampling and soil moisture detection have been compiled by Wilson (1980) and Everett et al. (1984).

The lysimeter is a capped tube with a permeable ceramic cup at the bottom through which the sample is collected. The ceramic cup is manufactured to a certain pore opening size to allow liquid entry. These samplers are inexpensive and versatile, and can be used up to several hundred feet deep. Oily or chemical contaminants tend to clog or alter the permeability of the ceramic, and may render it useless for sample collection. Another problem is that ceramic material may leach heavy metals from ceramic bulb or Teflon® cups and bias the sample. Lysimeters are relatively cheap to purchase and install. When carefully installed, they can provide chemical data from soil liquid that would otherwise be unavailable in the unsaturated zone.

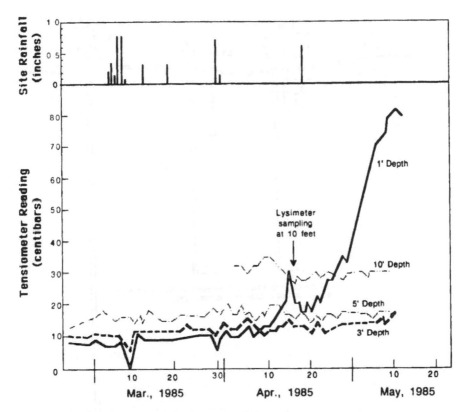

Figure 5   Vadose zone monitoring of precipatation penetration at a sewage sludge spreading area. Rainfall penetration into low moisture content sandy soil was monitored
with tensiometers and lysimeters for sludge leaching penetration above a perched
groundwater table. As the rainfall infiltrated, the tension decreased at depth,
showing a time lag of water infiltration. As tension fell, the moisture content was
increasing at the lysimeters, which were then sampled for vadose zone monitoring. (Modified from Merry and Palmer, 1985. With permission.)

Lysimeter installation usually involves collection of soil moisture data so an
estimate of available liquid can be made. Soil samples may also be collected to
ascertain indigenous chemistry or contaminant presence and soil moisture, which
help to calibrate conditions at the start of the monitoring program. In some cases,
tensiometers or other moisture-measuring devices are used to measure increased
moisture, a signal that soil liquid sampling should start. The lysimeter unit is
installed in a borehole with the ceramic tip encased in a fine-grained porous
material (such as very fine sandblasting glass bead), which forms a hydraulic
envelope about the cup to protect it from clogging. Once installed, a pressure is
applied to the unit that induces a suction at the ceramic, creating a pressure
"gradient" to overcome moisture surface tension and cause moisture movement
of the water films on soil particles. After a nominal time period (several to 24

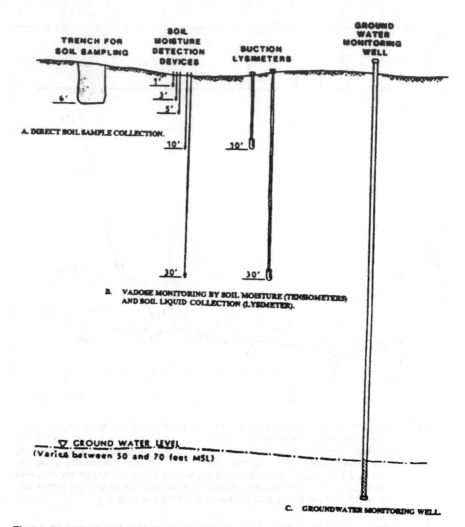

Figure 6 View of a combined unsaturated and saturated sampling array. (From Merry and Palmer, 1985. With permission.)

hours), the pressure is released and the sample that has accumulated in the tube is pumped to the surface and collected from the discharge line. Lysimeters can be used in low-moisture soils and can have long and useful service lives when properly installed and sampled (Merry and Palmer, 1985; see Figure 5). A conceptual example of combined monitoring using soil sampling, lysimeters and monitoring wells is present in Figure 6.

CHAPTER 6

Introduction to Regulatory and Legal Framework

INTRODUCTION

The field of ground-water quality and protection has become highly regulated over the past 20 years as environmental awareness grew in the scientific and general community. Hydrogeologists dealing with contaminant groundwater studies should have a good working knowledge of the local and state regulations pertaining to groundwater, potable water quality, and aquifer protection. Related areas involve consultant and client liability and ethical issues for hydrogeologists and other consultants. These issues are problems peripheral to the actual science of hydrogeology, but consultants should be aware of the possible effects of their investigations and field studies. As more regulations come on line, with different supervising agencies and hardening policy, confusion and redundant work can occur if a consultant lacks a good working knowledge of the pertinent rules.

Obviously, the legal issues would include book-length subjects in themselves. Legal and regulatory issues are covered and updated by books and seminars on a regular basis. The intent of this chapter is to give an overview of some applicable laws hydrogeologists may encounter and how they must function within the legal and regulatory boundaries. This chapter is meant to introduce readers to the regulatory world in the most general sense, assuming that they will have to address specific regulatory issues and be knowledgeable of related legal issues during investigations. Readers are referred to the Code of Federal Regulations and applicable state and local regulations for specifics, and must deal with these regulations on a case-by-case basis.

Existing laws have done much to order and direct environmental goals, but understanding them can be a bewildering task. The federal and state codes are the primary sources of the laws and regulations governing contamination. However, an excellent reference is Elliott (1988, and later dates), which presents a matrix of environmental laws with explanation of legislation by state. The regulations of several states (California, Florida, Illinois, New Jersey, Ohio, Pennsylvania, and Texas) have been compiled by Elliott and his associates, and the following discussion borrows heavily from that publication.

OVERVIEW OF FEDERAL LAW

Numerous laws and regulations have been passed in the 1970s and 1980s to deal with environmental issues and groundwater. Many federal laws, including the creation of the Environmental Protection Agency (EPA), have been passed that now govern or guide groundwater contamination investigations (see Table 1).

Table 1   Selected Federal Environmental Legislation

National Environmental Policy Act of 1970
Federal Water Pollution Control Act of 1972
Toxic Substances Control Act of 1976
Resource Conservation and Recovery Act of 1976 (RCRA)
Clean Water Act of 1977
Surface Mining Control and Reclamation Act of 1977
Safe Drinking Water Act of 1979
Comprehensive Environmental Resource Conservation and Liability Act of 1980
Hazardous and Solid Waste Amendments of 1984
Superfund Amendments Reauthorization Act of 1986

These laws established environmental protection and cleanup, as well as the definition of hazardous waste, hazardous waste transport, handling, cleanup and disposal. Laws were also added to address different aspects of environmental problems. Interestingly, there is no single federal groundwater protection statute (Patrick et al., 1987). While federal laws do provide for groundwater protection, they tend to focus on a narrow range of polluting activities (Patrick et al., 1987). The Safe Drinking Water Act of 1974 provided EPA authority to promulgate primary and secondary drinking water standards to public water supply systems. EPA's Office of Groundwater Protection released guidelines for groundwater classification in 1986. That system consists of three general groundwater classes; I — special groundwater; II — groundwater currently and potentially a source of drinking water; and III — groundwater not a source of drinking water. This system is based upon drinking water as the highest beneficial use of the resource.

A complete legal review is beyond the scope of this book, and readers are referred to the various applicable codes of regulations. However, to introduce this legal world, a highly generalized and brief review of some related federal and state regulations is presented since these are concerned with groundwater issues. It will also illustrate how complex the legal frameworks can become.

Resource Conservation and Recovery Act (RCRA) and Comprehensive Environmental Resource Conservation and Liability Act (CERCLA)

The RCRA law intent (summarized by Elliott, 1987) is to provide the "cradle-to-grave" regulation of hazardous wastes, and is actually constructed upon the Solid Waste Disposal Act of 1965, RCRA, the Hazardous and Solid Waste

Amendments (HSWA) of 1984 and part of the Superfund Amendments and Reauthorization (SARA) Act of 1986. The portions most often encountered by hydrogeologists are contained in the Code of Federal Regulations (CFR) 40 Part 264, Subpart F, which provide the geologic and hydrogeologic investigation approaches to siting and groundwater monitoring of hazardous waste disposal, storage, or treatment sites. These regulations outline the geologic information that federal regulators require for protection of aquifers underlying and near these sites. These also include seismic safety, engineering, geologic, and geotechnical engineering requirements, in addition to the groundwater aspects of facility siting.

Basically, the groundwater requirements should provide knowledge of the subsurface, installation of properly designed and constructed monitoring wells, and understanding of site-specific background groundwater geochemistry and contaminant data. In this way, the facility geology and its relation to nearby drinking-water sources can be utilized to protect those resources. A statistical analysis of site-monitoring information and indigenous baseline water quality is compiled and used to ascertain whether significant changes have occurred in water quality at detection monitoring points. If this occurs, then monitoring at a quality standard compliance point is established and monitored while the site is cleaned up. Once cleanup is finished or facility use ends, then a postclosure monitoring is performed for a considerable duration (up to 30 years) to ensure that the problem has been properly addressed. Needless to say, these are long, complicated, and highly costly investigations, requiring many staff and effort on the part of the consultant and client, and usually negotiation with the regulators. Investigations of military sites, old industrial facilities and landfills, and their cleanup action will continue for some time to come.

A related law is CERCLA, or "Superfund," which was created to identify any sites contaminated by release of hazardous materials and finance the remediation by "responsible parties" or from federal or state cleanup funds. The CERCLA Act of 1980 was amended by SARA, which strengthened Superfund. While EPA implements this program, some elements may be controlled by the states, which may become the lead agency for a particular cleanup.

Basically, sites are initially identified to EPA by owners and regulatory agencies, including those already identified under RCRA. Next, the Superfund sets priorities for cleanup under the National Contingency Plan using a rating system. The third step is to identify the owners and clean up the sites. The agencies seek to identify the potentially responsible parties (PRP) who can be required to pay for the remedial investigation/feasibility investigation to determine the extent of contamination and provide the bench test and design of the treatment and cleanup, and cleanup itself, directly or through reimbursement of federal expenditures. If no PRP is identified, then the Superfund would finance these activities.

EPA and State Underground Storage Tank Programs

RCRA and CERCLA are used for industrial, military, and civilian sites to clean up the contamination problems. However, another scale of contamination problems is now coming to the forefront: leaking underground storage tanks

(USTs) which occur everywhere in the United States, commonly the corner gasoline service station or small industrial site. Large contamination problems can arise from leaking tanks, whether the problem arises from large tank farms or one leak releasing a large quantity of contaminant, especially if classified hazardous. Consequently, many states have or will be implementing UST programs to deal with these problems, which can be from numerous diffused sources or a large point source. Some states, notably California, Florida, and New York, instituted UST programs in 1983–1984; most states are now responding to federal standards developed by EPA under HSWA requirements.

The UST approach differs somewhat from the RCRA approach in that the scale is usually smaller in investigative effort and sometimes financial costs; however, the level of care must be the same for both types of studies. If not, the smaller problem can easily grow to huge proportions together with legal and liability problems rivaling RCRA–CERCLA. It seems that the size of the overall UST problem is large and the tanks in small commercial yards, boiler fuel tanks, home motor fuel tanks, and farm tanks represent tens of thousands of dollars of potential problems. The potentially responsible parties (PRPs) for these sites are often people or organizations of limited financial resources and even a "small" contamination event may have a ruinous financial implication.

The EPA's UST program (40 CFR Parts 280 and 281) was passed to provide for regulation of underground tanks nationwide. This program will work with the state and local agencies to detect leaks, prevent spills, monitor tanks, and clean up subsurface leakage. In addition, the tanks must be properly closed and a financial responsibility section is included to address damage and cost of cleanup. Although many tanks apparently fall under this program, many others do not. These include farm and residential tanks smaller than 1100 gallons storing for noncommercial use, on-site heating oil tanks, tanks on or above the floor of underground areas, septic tanks for storm and wastewater, flow-through process tanks, tanks of 110 gallons or less, and emergency spill and overflow tanks.

Other sections of the regulations require overflow and corrosion prevention and detection, soil and groundwater monitoring, secondary containment of tanks, interstitial monitoring for chemical storage tanks, and notification of agencies. There is guidance for monitoring well placement, allowing some flexibility for well placement and design, especially for immiscible (floater) contaminants. As these regulations are implemented and evolve, changes, amendments, and policy evolution will occur as various programs deal with their respective problems.

SELECTED STATE AND LOCAL REGULATIONS

Several other states have passed UST laws and regulations in the past several years, including very extensive programs in New Jersey and Florida. California passed a comprehensive tank law in 1983, and several counties had passed underground tank ordinances prior to the statewide law. In all cases the intent in all states is similar; that is, to provide guidance on basic requirements for monitoring, construction of monitoring wells, soil and groundwater sampling, and periodic monitoring of the subsurface tanks and piping installation. Other regulations include hazardous materials management plans, UST and sump registration, permit fees for monitoring wells' installation and destruction, and cleanup equipment. These local regulations help identify the location of potentially hazardous materials for emergency planning, as well as both domestic and monitoring wells.

A newer wrinkle in the regulations is that permit fees and review of regulating agencies are now contributing to the financing of the agency work. In other words, the client may pay as he goes to permit for activities he must do on his site, including reviewing agency files and having agency personnel review and direct the specific case. Also local agencies may become the local implementing agency to direct the work if state or other agency personnel are unavailable. This problem of manpower and municipal funding to properly oversee the regulated community may slow agency action.

Example of Local Regulations Development

Discovery of leaks from subsurface solvent and fuel storage tanks in the "Silicon Valley" located in Santa Clara County at the southern end of the San Francisco Bay created concern about hazardous materials. The development of the hazardous materials storage ordinance (HMSO) was developed in Santa Clara County in 1982–1983 by a task force including participants from local government (Elliott and Esquibel, 1986). As discussed by Elliott and Esquibel, the Santa Clara Fire Chiefs' Association sponsored the effort because the firemen needed to know the location of aboveground hazardous material storage. Other groups joined the effort because they were concerned about leaks of material from USTs. In 1983, Santa Clara County and its 15 constituent cities began to implement HMSOs. As of 1989, there were 100 local UST programs in California, with 57 run mostly by county health agencies and 43 run in cities mostly through fire departments (Elliott, 1990). Federal UST regulations in place since 1989 add to the regulatory oversight and layered review.

The Santa Clara Valley Water District, a local water district, implemented UST groundwater monitoring guidance in 1983 in response to the numerous tank leaks that had occurred from both fuel and chemical storage. This ordinance provided for installing monitoring wells and instituting periodic monitoring for both water and soil vapor, and analyzing soil samples from the vicinity of the tank or tank complex. Soil and groundwater sampling site study protocols were established. Permits were required so that monitoring-well locations were public knowledge, as well as to review the basic construction specifications required by the ordinance. This documented the local site geology, and provided a preliminary estimate of soil and water quality and leakage extent. As the effort continued, cost-recovery measures were used to recover agency costs for implementation and case management.

Thus, layers of regulations may exist at one site: city or county, water district, state, and possibly federal, depending on the case. Hydrogeologists must therefore review the regulations and agencies with which they will be involved. This has a direct bearing on the completeness of the work required to address the needed guidance and reporting format. If the work is planned solely to address the specific regulations, then the work may be considered incomplete. If monitoring points are not sampled according to an accepted protocol, collected data could be deemed invalid. The possible confusion of the client and consultant is obvious with the overlapping review, and delegation of review to a local lead agency whose authority may lie in a state agency. It is always the responsibility of the supervising technical professional to be conversant with the laws and regulations pertaining to his client's location and site.

AGENCY INVOLVEMENT AND NEGOTIATION WITH THE CONSULTANT

The regulating agencies may not have the "answer" for the consulting hydrogeologist for regulatory gray areas. For example, many groundwater regulations exist, but soil contaminants in the vadose zone (above groundwater occurrence) are not regulated as specifically, or could fall into a mixed review (perhaps both county health and water district). Other contaminants may be involved for which regulated standards do not exist or toxicological data are incomplete. Therefore, while these contaminants may need to be removed, cleanup guidelines may utilize concentrations of contaminants or even other regulations not originally used for these particular cleanup purposes. At times soil samples may reveal contamination, but the meaning of the chemical concentrations is unclear, since transport regulations are used to determine if the constituent is present in soil or water at levels deemed to be hazardous and not as cleanup standards.

Agency guidance documents set forth sampling frequency, types of chemical analytical techniques, and the site-specific cleanup standard (how clean the site needs to be). For example, a site may be required to be cleaned to 1 part per billion chemical X and monitored for some duration on a quarterly basis. These would be site-specific requirements. Since approaches to regulation application are fluid, and the toxicological knowledge base (including contaminant transport) is growing, final disposition cleanup goals may change. However, this difficulty may be eased if the consultant has a working knowledge of the applicable regulations, has conducted a complete and thorough investigation, and has experience negotiating with clients and agencies.

Because the consultant is dealing with laws and supervising agencies, the investigation paper flow is sent to these people. The consulting hydrogeologist, acting on the client's behalf, presents site information and plan development to the regulating agencies. The consultant may have to send written plans, rationales, sampling protocol, periodic monitoring reports, and other documents for agency review to support the case and interpretation of the collected data. This includes meetings with the case officer assigned to the site. In turn, this involves negotiation of investigation approaches, data validity, monitoring system development, and ultimate site cleanup plans. The subsequent remediation plan negotiation is predictably difficult, since the extent of cleanup will dictate how much money will be required for cleanup. Obviously, the "how clean is clean" debate can become vigorous. Because the agency will decide how far the cleanup will proceed, and set the chemical concentration cleanup levels, the consultant's input mainly involves the site hydrogeochemistry. The "burden of proof" will rest with the client and consultant, so the regulator must understand the situation well enough to establish the cleanup goal that is realistic for the site. These are difficult problems to solve and all parties desire the same end — environmental compliance and protection.

Examples of Regulator Negotiation and Interaction

A growing field of environmental work is "reconnaissance site assessments," sometimes called "environmental due diligence." These types of investigations are done to initially ascertain whether a site has an environmental problem (in this case, subsurface contamination investigation). Since property owners, banks, or other involved parties may become liable for environmental compliance or cleanup, these studies are increasingly involving the site history and contaminant potential. The need for future environmental compliance may or may not be required by a regulating agency regardless of property ownership. In order to avoid potential liability, clients will need to prove that they did not cause the problem or that the problem came from an off-site source.*

Palmer and Elliott (1988) have suggested a possible approach for this type of problem. When working with the agencies, the consulting hydrogeologist should supply the needed information to the regulators so they can evaluate site conditions. If some information is not forwarded to them, or work was incomplete, the position you present may not be acceptable to them. All records, field data, and chemical data should be completely documented, and copies filed with the agencies so a paper trail exists in their files. Unreasonable demands and confrontational meetings tend only to polarize individuals and hinder progress for both sides. These projects will have substantial costs in time and money, regardless of size. The best approach is to have the material in hand, together with historical files that document and support your position. Good lines of communication with regulators are needed to accurately state the site conditions and needs in the context of applicable regulations. Negotiation should be firm but amicable. Two case histories are given from Palmer and Elliott (1988) to illustrate this approach:

Case 1

A convenience store possesses on-site underground gasoline storage tanks which are monitored by UST leak detectors and one monitoring well. A sudden appearance of gasoline in his single monitoring well indicates product is moving under the site. The tank leak detectors show no leak. The owner elects to install two more monitoring wells on his upgradient property line and discovers gasoline

* A related first step is the preliminary site assessment, which is usually a site walk to document existing site conditions, and review existing regulatory files and databases to search for site problems; typically subsurface invasive testing is not done, or is very limited.

migrating on-site from an upgradient source. He monitors this quarterly. Six months later, State and local agencies inform the owner and several other nearby subsurface tank owners may be named as responsible parties for a large gasoline plume in the vicinity.

Our store owner sent the monitoring well installation reports and quarterly monitoring to the State so they were in State files months before the State started looking for responsible parties. This evidence of two upgradient wells, and quarterly monitoring of all three wells show he is not responsible. The State does not name our owner as a responsible party, but does request that he share monitoring information.

By installing two wells and keeping abreast of the problem, the store owner saved himself from being entangled in the responsible party battle, and cooperated with the agencies. At a cost of about $12,000 (1988) dollars over one and one-half years, he saved himself from being ordered to do a $50,000+ investigation over a three month period — and the future grief from being involved in a cleanup problem which he did not create.

Case 2

A development company (buyer) wants to buy a portion of a valuable property. While the site does not appear to have had hazardous materials on it, the buyer elects to perform an environmental reconnaissance investigation by installing monitoring wells and collecting soil and groundwater samples for chemical analysis. Samples from the wells reveal slight contamination beneath the site by industrial solvents, in concentrations which exceed State Action Levels. Monitoring wells across the property line (installed by another consultant for the owner (seller) confirm the contamination, but it turns out that owner does not have a history of use of these chemicals. The State agency would want "proof" that the contamination was not caused by the current owner since the State surmised a history of solvent use on or near the subject property. Otherwise the State may want the current or future owners to become involved in a future investigation and cleanup.

Since two consultants are involved, one for the buyer and one for the seller, additional investigation is done to verify that contaminants exist, confirm which way water is moving, and localize history of spills in the region. The additional surficial soil sampling show that contaminants are not present and so could not have migrated through the vadose zone to groundwater. This eliminates an on-site source. A record review show several large industrial solvent spills had occurred upgradient and were moving in the direction of the property. The limited data from monitoring wells and other publicly available reports on those large spills indicate that the plumes may have arrived at the site, or the plume edge was in the vicinity of the property.

Here the State does not render a "final" option as to responsibility, but can require future monitoring work. But the documentation gathered and shared by the two consultants, collected from numerous sources, shows that: a) the site soil was not the source of the contaminants, b) large upgradient solvent plumes had occurred and were moving toward the subject property, and c) the additional investigation confirmed the initial presence of solvents and sources were traced, if not to a point of origin, at least to several possible sources, which did not involve the current owner or prospective buyer. The approach has been to collect the data and construct an argument which takes into account the types of information which the State would request to evaluate contaminant sources *as if* an on-site investigation were required. The work performed should meet the requirements of the State if a future challenge is made. Since the sources of contamination are traceable to some off-site point, the likelihood of being named in a cleanup seems remote, but not certain. Both owner and potential buyer now must resolve potential risks and fine points of the sale given current hydrogeochemical knowledge.

EXPERT REVIEW

At times there may be a need for a consultant to review existing investigation documents to ascertain whether the PRP is in fact responsible for the contamination problem. Consulting firms as a whole may be capable of the study, but at times additional expertise may be desirable. The expert is someone with a recognized track record in groundwater investigation experience (just as an expert in any other field would have outstanding credentials; at times there are "experts" who have alleged experience and credentials; caveat emptor for PRPs). The person(s) or firm(s) may be retained directly or assist the on-board consultant for review of the work, and possibly when the project enters litigation. Testimony and trial experience of the expert may be as important as the expertise in some instances. Since numerous sites have entered litigation in the past, and the financial responsibility for cleanup can be immense, the opposing experts are called in to render their opinions on the hydrogeologic technical aspects, cleanup costs, and other salient points on the case.

The expert will review the information and render an opinion, which may be pivotal in deciding who possibly caused the problem and who pays. As in other arenas, hydrogeologic experts would exhibit their experience, and the possible weaknesses of the opposing consultant's argument, to the client's benefit. Sometimes the expert review may be backed by site information, or the lack of specific information to support the opponent's position. For example, a preliminary or reconnaissance study may not have a degree of completeness to support a relatively definitive conclusion. Years of study at another site, defining contamination position, timing and extent, chemical signatures, groundwater motion, plume mapping, and so on could be difficult to pick apart in refuting that site's sources. But, an intense sampling at only one location while ignoring other possible site sources may be suspect in identifying "*the* source" for which that single PRP is responsible. An expert may skew the facts selectively with creative interpretive answers, and overly selective interpretation may fade when shown in light of all the data. The opinion rendered should be realistic given the documentation and realistically applied to the site hydrogeology. The expert's greatest service to clients is to realistically help them know their position and exposure.

SUMMARY

The consultant must have a good working knowledge of federal, state, and local regulations regarding the site consideration. This is basic for completing investigations that comply with the guidance and provide the necessary monitoring and cleanup approach. At times, regulation layering may require interaction with different agencies. The "burden of proof" will usually always be on the client (and the consultant) to show the extent of contamination and negotiate remediation. Typically, negotiation with agency people will occur, and complete and up-to-date information is needed for agency review. Negotiation should be firm and amicable for the work necessary to get site-specific answers. The consultant should obtain information that anticipates regulator questions, provides answers within the context of applicable regulations, and provides the client with a realistic problem solution.

CHAPTER 7

Introduction to General Groundwater Geochemistry

INTRODUCTION

This chapter introduces geochemistry and laboratory analysis as related to applied contaminant hydrogeology. A vast body of literature exists on groundwater geochemistry, a thorough review of which is beyond the purpose of this chapter. This reviews the soil and groundwater quality chemical considerations of a site, because the two are interrelated in terms of contaminant migration and regulatory reporting. Any site has an indigenous natural geochemistry that has to be quantified at the beginning of the study. Also, the investigator will have to evaluate the anthropogenic contaminant geochemistry. Groundwater quality for the purpose of the following discussion involves two arenas: first, general geochemical parameters that have regulated criteria for drinking water, and for the protection of drinking water aquifers; and second, an outline of laboratory selection and analyses.

Groundwater has a geochemical variability caused by natural processes. These include groundwater flow, the formations through which flow occurs, source of sediments that compose the aquifers, chemical changes resulting from annual flow fluctuations, recharge sources, and mixing with other groundwater of differing chemistry. In a natural system elements enter or leave the system, or form compounds within it (see Figure 1; Toth, 1984). Groundwater quality depends on the substances dissolved in the water and the chemical behavior of the compounds in the water imparted by the indigenous geology through which it flows. Consequently, groundwater quality is changeable as it moves through the aquifer. Changes may also be imparted, albeit very slowly, from leakage through aquitards or aquicludes (Toth, 1984).

It is important to quantify the site groundwater geochemistry under consideration so that the baseline quality is established at the beginning of the study. This includes both the soil and groundwater, since recharge through the soil or

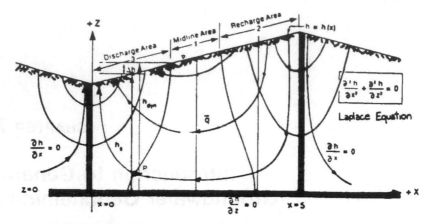

The Unit Ground Water Basin. geometry, boundary conditions, unit patterns of hydraulic head h and flow q̄ and areas of the three basic ground water flow regimes

Ground Water Chemistry and Hydraulic Regimes of Flow Systems

Recharge areas

Conditions: rain water; low TDS; high CO_2; low T; gradp−q̄; cross formational q̄

Processes: dissolution, hydration, oxidation; attack by acids; base exchange

Consequences: Dominant species — Ca, Mg, HCO_3, CO_3, SO_4
diverse rock types — diverse constituents
rapid increase in TDS

Midline areas

Conditions: source water moderately changed; p~ hydrostatic; ρ̇ ~ 0; T ~ const; low free CO_2; little cross formational q̄

Processes: dissolution, precipitation, sulfate reduction, base exchange

Consequences: Dominant species — Na, Ca, Mg, HCO_3, SO_4, Cl; gradual increase in TDS

Discharge areas

Conditions: highly mineralized source water gradp and gradT opposite to q̄; cross formational flow; mixing with descending fresh water.

Processes: precipitation; reduction; membrane filtration

Consequences: high TDS possibly decreasing upward; dominant species — SO_4, Cl, Na

General Changes in Direction of Flow

TDS : Increase

$\frac{SO_4}{Cl}$: Decrease (SO_4 reduction; higher solubility of Cl)

$\frac{SO_4}{HCO_3}$: Increase (depletion of CO_2)

$\frac{Ca}{Na}$: Decrease (No Ca added due to depletion of CO_2; exchange of Ca for Na)

$\frac{Ca}{Mg}$: Decrease (no Ca added; Mg SO_4 more soluble than Ca SO_4)

Figure 1 The Unit Groundwater Basin and natural geochemical changes as groundwater moves through the system. (From Toth, 1984. With permission of the National Ground Water Association.)

sediment cover can chemically impact aquifer water. Groundwater moves beneath the site and water quality can change through time, whereas the soil or sediment chemical "quality" generally will not change. However, water moving vertically through the vadose zone from surface contaminated sites may significantly affect and change groundwater on arrival.

Developments in chemical analytical techniques and regulatory criteria affect the investigation, since standards may change during the course of study. Analytical techniques have become more refined over the last 10 years, so resolution to parts per million (ppm) and parts per billion (ppb) and lower is now possible. Thus, a constituent that was not detected at the ppm range may now be detected, and be of concern, at the ppb range. Because toxicological studies determine chemical health risks at certain concentrations, these concentrations become critical to the investigation in that they may or may not change the regulated standard. Regulations may change and their interpretation could affect the course of the investigation, and will probably increase with the use of risk assessment for site evaluation and closure. Consequently, when a health risk is determined, it is likely that the regulatory interpretation will become more conservative. If the regulated concentration of the contaminant is lowered, the cost and time of the investigation and cleanup will probably increase.

INORGANIC COMPOSITION AND QUALITY

Inorganic groundwater chemistry deals with the physical and chemical factors that govern groundwater movement. Detailed discussions of inorganic chemistry of groundwater may be found in Hem (1985) and Matthess (1982) (see Table 1). Usually four criteria are used when the monitoring well is sampled. These chemical parameters yield quick and inexpensive chemical information and allowing the hydrogeologist to observe site groundwater geochemistry at the time of well sampling. The U. S. Geological Survey has standardized values of these parameters for general geochemical classification.

Electrical Conductivity

Electrical conductivity (EC) refers to the ability of a substance to conduct electrical current (Hem, 1985). The ability to transmit an electrical current depends on the concentration of charged, or ionic species in the water. Hence, the measure of the conductance is used to approximate the total concentrations of ionic species present, and a rough approximation of how mineralized the water. Measurement of electrical conductance is usually in micromhos per centimeter (μmhos/cm) or siemens per centimeter (S/cm). Usually the measurements are standardized to 25°C. Since numerous influences may cause EC measurements to change, it is only a gross estimator of dissolved salt load or contamination. Field EC measurement equipment available today commonly utilize self-calibrating adjustments, or are calibrated to known standards when they are used.

pH

pH is the measure of the alkalinity and acidity of the groundwater, or hydrogen concentration on a logarithmically calculated scale where 1 (most acidic) to 14 (most basic or alkaline) pH units (Gymer, 1973). pH has considerable influence

Table 1A   Inorganic Substances Affecting Groundwater Quality and Use

Substance	Major natural sources	Effect on water use	Concentrations of significance (mg/L[1])
Bicarbonate (HCO$_2$) and carbonate (CO$_3$)	Products of the solution of carbonate rocks, mainly limestone (CaCO$_3$) and dolomite (CaMgCO$_3$), by water containing carbon dioxide.	Control the capacity of water to neutralize strong acids. Bicarbonates of calcium and magnesium decompose in steam boilers and water heaters to form scale and release corrosive carbon dioxide gas. In combination with calcium and magnesium, cause carbonate hardness.	150–200
Calcium (Ca) and magnesium (Mg)	Soils and rocks containing limestone, dolomite, and gypsum (CaSo$_4$). Small amounts from igneous and metamorphic rocks.	Principal cause of hardness and ofr boiler scale and deposits in hot-water heaters.	25–50
Chloride (Cl)	In inland areas, primarily from seawater trapped in sediments at time of deposition; sition; in coastal areas, from seawater in contact with freshwater in productive aquifers.	In large amounts, increases corrosiveness of water and, in combination with sodium, gives water a salty taste.	250
Fluoride (F)	Both sedimentary and igneous rocks. Not widespread in occurrence.	In certain concentrations, reduces tooth decay; at higher concentrations, causes mottling of tooth enamel.	0.7–1.2[2]
Iron (Fe) and manganese (Mn)	Iron present in most soils and rocks; manganese less widely distributed.	Stain laundry and are opjectionable in food processing, dyeing, bleaching, ice manufacturing, brewing and certain other industrial processes.	Fe>0.3, Mn>0.05
Sodium (Na)	Same as for chloride. In some sedimentary rocks, a few hundred milligrams per liter may occur in freshwater as a result of exchange of dissolved calcium and magnesium for sodium in the aquifer materials.	See chloride. In large concentrations, may affect persons with cardiac difficulties, hypertension, and certain other medical conditions. Depending on the concentrations of calcium and magnesium also present in the water, sodium may be detrimental to certain irrigated crops.	69 (irrigation), 20–170 (health)[3]
Sulfate (SO$_4$)	Gypsum, pyrite (FeS), and other rocks containing sulfur (S) compounds.	In certain concentrations, gives water a bitter taste and, at higher concentrations, has a laxative effect. In combination with calcium, forms a hard calcium carbonate scale in steam boilers.	300–400 (taste), 600–1,000 (laxative)

[1] A range in concentration is intended to indicate the general level at which the effect on water use might become significant.
[2] Optimum range determined by the U.S. Public Health Service, depending on water intake.
[3] Lower concentration applies to drinking water for persons on a strict diet; higher concentration is for those on a moderate diet.

Table 1B   Characteristics of Water that Affect Water Quality

Characteristic	Principal cause	Significance	Remarks
Hardness	Calcium and magnesium dissolved in the water.	Calcium and magnesium combine with soap to form an insoluble precipitate (curd) and thus hamper the formation of a lather. Hardness also affects the suitability of water for use in the texile and paper industries and certain others and in steam boilers and water heaters.	USGS classification of hardness (mg/L as CaCO$_3$) 0–60: Soft 61–120: Moderately hard 121–180: Hard More than 180: Very hard
pH (or hydrogen-ion activity)	Dissociation of water molecules and of acids and bases dissolved in water.	The pH of water is a measure of its reactive characteristics. Low values of pH, particularly below pH 4, indicate a corrosive water that will tend to dissolve metals and other substances that it contacts. High values of pH, particularly above pH 8.5, indicate an alkaline water that, on heating, will tend to form scale. The pH significantly affects the treatment and use of water.	pH values; less than 7, water is acidic; value of 7, water is neutral; more than 7, water is basic.
Specific electrical conductance	Substances that form ions when dissolved in water.	Most substances dissolved in water dissociate into ions that can conduct an electrical current. Consequently, specific electrical conductance is a valuable indicator of the amount of material dissolved in water. The larger the conductance, the more mineralized the water.	Conductance values indicate the electrical conductivity, in micromhos, of 1 cm^3 of water at a temperature of 25°C.
Total dissolved solids	Mineral substances dissolved in water.	Total dissolved solids is a measure of the total amount of minerals dissolved in water and is, therefore, a very useful parameter in the evaluation of water quality. Water containing less than 500 mg/L is preferred for domestic use and for many industrial processes.	USGS classification of water based on dissolved solids (mg/L): Less than 1,000; Fresh 1,000–3,000; Slightly saline 3,000–10,000; Moderately saline 10,000–35,000; Very saline More than 35,000; Briny

From Heath, 1982. With permission.

on the water geochemistry because it affects ionic strength, oxidation–reduction and organic carbon content, and the mobility of metallic ions. This measurement is considered to be more accurate in the field because pH may change due to temperature changes, carbon dioxide or other gases escaping, or other gas entry (Gillham, 1983). For example, when sampling a stable reading of pH during water extraction, typically it is accepted as the true pH of aquifer water, and is noted prior to the sample collection.

Oxidation-Reduction

This is also known as the "redox" potential (Eh). This is the measure of the relative intensity of oxidizing or reducing conditions in solutions provided by the Nerst equation (Hem, 1985). When the pH is known the stability of minerals in water can be determined. Measurements of dissolved oxygen may indicate whether or not groundwater has an oxidizing condition (API, 1983).

Total Dissolved Solids

Total dissolved solids comprise dissociated and undissociated substances in the water (Matthess, 1982). The value is commonly determined by evaporating a water sample to dryness, although the residue is slightly different from the solution due to minor losses and precipitation. This is a common indicator of overall quality, and water containing less than 500 mg/l is preferred for domestic and industrial use (Heath, 1982).

REGULATIONS ESTABLISHING DRINKING WATER QUALITY STANDARDS

Standards have been set forth by the federal and state governments for the minimum drinking water quality for human consumption. The Federal Office of Drinking Water has established Recommended Maximum Contaminant Levels (RMCLs) for contaminants in water. The RMCLs have been developed from and are health-based standards derived from toxicological data. RMCLs are health-related goals and are not enforceable drinking water standards. However, the Federal Primary Drinking Water Standards have established Maximum Contaminant Levels (MCLs), which are federally enforced. The MCLs are set as close as possible to the RMCLs after taking into account technology available and cost to achieve or meet the drinking water standard. Individual states have adopted either the federal criterion, or a modification of it, as their water quality standard. These standards cover some of the potential contaminants that may affect water supplies (see Tables 2 and 3).

Additional water quality information is available from the federal and state government sources regarding contaminants not covered under the aforementioned standards. Health advisories have been issued on certain chemicals that are derived from National Academy of Sciences and EPA information. The EPA

Table 2   National Interim Drinking Water Standards

Maximum Contaminant Levels for Inorganic Chemicals

Contaminant	Level, milligrams per liter (micrograms per liter in parentheses)
Arsenic	0.05 (50 μg/l)
Barium	1. (1000 μg/l)
Cadium	0.010 (10 μg/l)
Chromium	0.05 (50 μg/l)
Fluoride	2.2
Lead	0.05 (50 μg/l)
Mercury	0.002 (2 μg/l)
Nitrate (as N)	10.
Selenium	0.01 (10 μg/l)
Silver	0.05 (50 μg/l)

Standard	Milligrams per liter (micrograms per liter in parentheses) except as noted
Chloride	250
Color	15 units
Copper	1.0 (1000 μg/l)
Corrosivity	Noncorrosive
Foaming agents	0.5
MBAS (methylene-blue active substances)	
Hydrogen sulfide	not detectable
Iron	0.3
Manganese	0.05 (50 μg/l)
Odor	3 (Threshold No.)
Sulfate	250
Total residue	500
Zinc	5 (5000 μg/l)

Maximum Contaminant Levels for Organic Chemicals

Contaminant	Level, milligrams per liter
(1) Chlorinated hydrocarbons: Endrin (1,2,3,4,10, 10-hexachloro-6,7 expoxy-	0.0002

Maximum Contaminant Levels for Radium-226, Radium-228, and Gross Alpha Particle Radioactivity

(1) Combined radium-226 and radium-228-5 pCi/L
(2) Gross alpha particle activity (including radium-226 but excluding radon and uranium)-15 pCi/L

Radionuclide	Critical oxygen	pCi per liter
Tritium	Total body	20,000
Strontium-90	Bone marrow	8

has established the National Ambient Water Quality Criteria (NAWQC) under the authority of the Clean Water Act (1974). NAWQC are not mandatory standards but states can adopt them as enforceable standards to protect beneficial uses of water bodies.

NATURAL CONTAMINANTS

Although groundwater can be drinkable directly from the subsurface, it is not necessarily of drinkable quality everywhere, and in fact may need some kind of

Table 3   Contaminants Regulated Under Safe Drinking Water Act,
          1986 Amendments

Volatile organic chemicals	Organics
Trichloroethylene[a]	Endrin
Tetrachloroethylene	Lindane[a]
Carbon tetrachloride[a]	Methoxychlor[a]
1,1,1-Trichloroethane[a]	Toxaphene[a]
1,2-Dichloroethane[a]	2,4-D[a]
Vinyl chloride[a]	2,4,5-TP[a]
Methylene chloride	Aldicarb[a]
Benzene[a]	Chlordane[a]
Chlorobenzene[a]	Dalapon
Dichlorobenzene(s)[a]	Diquat
Trichlorobenzene(s)[a]	Endothall
1,1-Dichloroethylene[a]	Glyphosphate
trans-1,2-Dichloroethylene[a]	Carbofuran[a]
cis-1,2-Dichloroethylene[a]	Alachlor[a]
Microbiology and turbidity	Epichlorohydrin[a]
Total coliforms[a]	Toluene[a]
Turbidity[a]	Adipates
Giardia lamblia[a]	2,3,7,8-TCDD (Dioxin)
Viruses[a]	1,1,2-Trichloroethane
Standard plate count	Vydate
Legionella	Simazine
Inorganics	Polynuclear aromatic hydrocarbons (PAHs)
Arsenic[a]	Polychlorinated biphenyls (PCBs)
Barium[a]	Atrazine
Cadmium[a]	Phthalates
Chromium[a]	Acrylamide[a]
Lead[a]	Dibromochloropropane (DBCP)[a]
Mercury[a]	1,2-Dichloropropane[a]
Nitrate[a]	Pentachlorophenol[a]
Selenium[a]	Pichloram
Silver	Dinoseb
Fluoride[a]	Ethylene dibromide[a]
Aluminum	Dibromomethane
Antimony	Xylene[a]
Molybdenum	Hexachlorocyclopentadiene
Asbestos[a]	Radionuclides
Sulfate	Radium 226 and 228
Copper[a]	Beta particle and photon radioactivity
Vanadium	
Sodium	Uranium
Nickel	Gross alpha particle activity
Zinc	Radon
Thallium	
Beryllium	
Cyanide	

[a] Included in EPA proposed and final rules published in Federal Register, Nov. 13, 1985.

treatment in most areas today prior to distribution for potable use. Natural ground-water contaminants can include natural sources of petroleum, salts, trace elements, and biologic sources. Although the water may be usable, if it exceeds the (current state or federal) regulated chemical constituents, then it can be considered "contaminated" and unfit for human consumption. Regional geologic influence may have profound trace element geochemical effects on water quality. Since the observation of inorganic contaminants in water (and soil) may initiate a regulatory action, identifying naturally occurring contaminant sources is very important (see Table 3).

It is easy to see that trace element ranges may vary widely in any area depending on the local geology. For example, if an urban area is adjacent to a large body of ore-bearing rock, the eroded and decomposing rock yields trace elements (commonly called heavy metals) that may be released as transported

Table 4 Content of Various Trace Elements in Soils

Trace elements (metals)	Selected average for soils (mg/kg)	Common range for soils (mg/kg)
Al	71,000	10,000–300,000
Fe	38,000	7,000–550,000
Mn	600	20–3,000
Cu	30	2–100
Cr	100	1–1,000
Cd	0.06	0.01–0.70
Zn	50	10–300
As	5	1.0–50
Se	0.3	0.1–2
Ni	40	5–500
Ag	0.05	0.01–5
Pb	10	2–200
Hg	0.03	0.01–0.3

From Lindsay, 1979.

sediment, and ultimately enter the groundwater via solution. If a metal is revealed in a drinking water chemical analysis, then the water could appear contaminated if present above the regulated standard. Since similar metals are used in local manufacturing processes, is the observed source natural or anthropogenic? The solution may be simple if industrial development has not occurred upgradient, but if urban sites are sold for redevelopment, a check for potential contamination could cause problems unless evidence is available to show it was not human-derived.

BIOLOGIC CONTAMINANTS

Biologic contaminants are very prevalent in groundwater, especially shallow groundwater; they can be very serious and are worth mentioning (see Mathewson, 1979; Miller, 1980). Animal and human wastes, leakage from sewers, and sewage treatment are primary urban and agricultural sources for these contaminants. The problems of disease from these sources are well known and are one of the initial regulatory concerns for water treatment, quality, and drinking water well-sealing standards, as enforced by county health agencies and plumbing codes. Wellhead testing or water-distribution testing for coliform bacteria is typically a weekly test for water purveyors. Other biologic sources may include nitrate loading into regional shallow aquifers from domestic leachfields in less urbanized areas and aftereffects of large-scale surface flooding fouling water supplies. A recent study by Robertson and Blowes (1995) shows that an acidic leachate may arise from septic system loading in leachfields and mobilize trace elements in carbonate poor terrain. Here, both biologic and metals contaminants could arise from a residential source as a groundwater threat.

ORGANIC CHEMICAL CONTAMINANTS

The quality of groundwater in terms of "organic quality" constituents for this discussion relates to the presence of manmade organic contaminants. Groundwater may contain some natural organic compounds. Organic carbon, humic, and fulvic acids are naturally present in groundwater and arise from organic decomposition and other inorganic processes (Dragun, 1988). Organic carbon concentrations are generally low in groundwater; due to long residence times the carbon oxidizes to carbon dioxide, contributes to alkalinity, or may recombine to form methane and can be adsorbed onto aquifer material. Petroleum-producing regions may reveal background trace concentrations of petroleum hydrocarbons in both soil and groundwater, which could appear to be refined-fuel-derived. For the purpose of the following discussion however, we will assume that has only groundwater natural inorganic components, and that organic materials found in groundwater have arisen from some anthropogenic contaminant source.

Information Sources for Organic Contaminants

The growth of chemical industry has involved both refined natural petroleum compounds and synthetic chemicals and materials, and waste by-products of those processes. Consequently, there are literally thousands of compounds that may be released into the environment and contaminate soil and groundwater (see Table 3). Organic contaminants may remain from long-ceased operations, such as coal-tar plants of the last century. Polychlorinated biphenyls (PCBs) have been used commonly most of this century, and may be found almost anywhere through use of solvents, pesticides, chemical sealers, plastics, explosives, munitions, and rocket fuels. Organic contaminant history from processing or manufacturing may be complex and extend far into the past. EPA has prepared a list of the contaminants of most concern, commonly referred to as the Appendix IX list or Safe Drinking Water Act. Typically, a history of the site use of chemicals is a first investigation step since the ownership and chemical uses and processes may vary widely over time.

Information on subsurface fate, transport, and behavior in the groundwater environment is somewhat limited since organic contamination has been recognized as a threat to environmental quality only in the past two to three decades. Government, manufacturers', and university research into the chemistry of compounds, their use, and their safety is ongoing. Research concerning fate and transport models are also ongoing, and more studies and results are now generally available through journals, site cleanup reports, and training seminars. To date, only a few contaminants (i.e., motor fuel hydrocarbons, pesticides, and halogenated solvents) have been studied extensively out of the thousands of potential contaminants. Given the complexity and plethora of industrial chemicals, the services of a chemical engineer and analytical chemist familiar with the chemicals, their properties, and analytical techniques are required.

Types of Organic Contaminants

Obviously almost any chemical could become a potential contaminant under the right circumstances. Sources of general chemical information should be

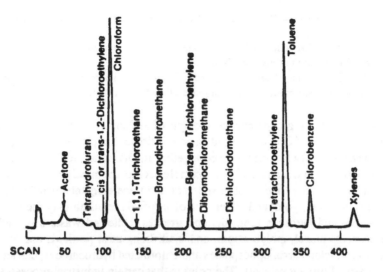

Figure 2 Example of a gas chromatrgraph-mass spectrometer chromatogram. (From Trussell and DeBoer, 1983. With permission.)

handy, and the reader is referred to chemical indexes and dictionaries regarding specific compounds (Lewis, 1993; Montgomery and Welkom, 1989; Sax and Lewis, 1987). The following contaminants are listed by groupings of EPA laboratory analytical methods, taken from the EPA Methods of Analyses for Water and Wastewater. These groups bring together contaminants having similar chemical properties, and are common in industrial use. This allows a hydrogeologist to identify initially the type of organic compounds present and to attempt to predict their gross movement and behavior in groundwater. Often in these analyses, groups of compounds may be identified in an EPA Method scan. An example of gas chromatograph-mass spectrometry chromatogram identification of volatile organic chemical group is presented in Figure 2. The following list is not meant to be comprehensive, but rather to name the general chemical groups that usually are of concern regulating agencies.

1. Volatile Organic Compounds — Typically these are compounds with low vapor pressure, and may include solvents, fuels with an aromatic chemistry (benzene ring type). It may include halogenated compounds (containing chlorine, fluorine, or bromine) such as chlorinated solvents and materials (trichloroethylene, methylene chloride, or chloroform). Depending on the compound, miscibility and solubility vary.

2. Acid-Base-Neutral Compounds — These may include polynuclear aromatics, ethers, esters, phenols, PCBs, plasticizers, and similar industrial compounds.

3. Agricultural Chemicals — These include the vast array of pesticides, herbicides, nematodicides, fertilizers, and related chemicals. These can be significant given the quantity of chemicals used in modern agriculture, which are applied directly to the ground. Some older compounds such as DDT are long-lived, but migrate conservatively, whereas others such as DBCP may be mobile in the groundwater environment.

4. Trace Elements — This includes selected trace elements (13 metals), asbestos, and cyanides.
5. Alcohols and Ketones — These are common as cleaners and may tend to move rapidly due to their high solubility.
6. Oils, Greases, and "Heavy" Petroleum Products — These cover a wide range of hydrocarbon materials that compose other fuels and lubricants.

Sometimes these groups may be thought of as "contaminant suites" of compounds, or groups of chemicals used in certain industrial processes of manufacture. Since these may at times be used together in industrial processes, they may also be possible contaminants. For example, some chemicals associated with electronic manufacture are volatile solvents and trace elements arsenic, copper, lead, zinc, mercury, gold, and silver. Petroleum-fuels retailing may include gasoline, diesel and kerosene fuel, oils and greases, collected waste oils, and the trace elements nickel, molybdenum, and lead. Long-term pesticide use or manufacture can include organophosphates and chlorinated hydrocarbons, carrier oils, and carriers of mixing solvents. The point is that certain industrial processes can include similar types of chemicals, and those chemicals could become threats to soil and groundwater. Initial chemical testing may need to include groups of chemicals, and the analytical scans often reveal the groups with similar properties in the tests, or those amended to the specific compound. Certain chemicals may not have a "standard EPA Method" and specialty tests need to be performed as needed for specific chemicals.

For an example of this, consider that investigation of petroleum spills and contamination comprises a large portion of soil and groundwater problems. EPA methods of analysis can give the overview of the identification, although data must be carefully reviewed before accepting the "truth" of the data. Often the contaminants are thought to be simply "gasoline" or "diesel" where the actual identification process is not necessarily simple. Since fuels may contain numerous compounds, and these can change through the initial manufacturing process, and degrade significantly in the subsurface, care must be taken to properly identify them. Zemo et al. (1995) discuss "fingerprinting" these compounds to take into account their composition, refining, method of analysis, and interpretation of the data. Proper identification may preclude the need to continue investigation and cleanup action, or avoid establishing an inappropriate site cleanup goal. Further, some degree of degradation may be inferred from chromatograms that can cause misidentification of the fuel and its components.

ANALYTICAL LABORATORIES

The selection of the analytical laboratory used through the investigation is a critical decision. Once chosen, the laboratory will generate all the chemical data the consultant reviews and reports to the regulating agency. The accuracy, validity, and reproducibility of the laboratory results are vital to the success of any study because they provide the "direct" chemical observational data. The laboratory should be able to perform all the analyses required by the investigation (such as chromatography, mass spectrometry, atomic adsorption, etc.; see Figure 2). Some contract laboratories may have specialty analyses, such as for pesticides, or may not do certain analyses. If the analyses and results are suspect, the data may be discounted and the concentration or even presence of the contaminant may be questioned. It also may create litigation problems if errors are encountered in methods, analysis, quality control, and chain-of-custody documentation.

Contract analytical laboratories usually perform analyses according to recognized standards. These include American Society of Testing Methods (ASTM) and EPA standard methods that are reviewed by government and related professional societies and industrial research facilities. Laboratories may be certified by state agencies, whereas other states need only have laboratories attest to the fact that they use approved EPA methods. If a government certification is required by the overseeing agency, then the laboratory must have the certificate prior to doing the analysis. As noted above, some compounds of interest to the investigation may not have "standard" methods of analysis. Some methods may be compound-specific, as for certain pesticides or oils. Analytical methods may need to be modified depending on the type of contaminant and laboratory capability. Hence, it is advantageous to contact the laboratory director to discuss what work and analyses are desired prior to submitting samples for analysis. Over time, the hydrogeologist may develop a relationship with the chemist performing the analysis, and learn some of the intricacies and the limits of the technique.

Laboratory Selection Criteria

A general procedure for laboratory selection has been evolved by EPA. The following criteria summary is drawn from the EPA Groundwater Technical Enforcement Guidance Document (1986, revised 1992). The quality assurance and control procedures used by the laboratory must be thoroughly reviewed. Note that individual states may have guidance documents or criteria that may amend certain analyses referenced for a particular investigation. Additional guidance documents and material prepared by specific laboratories should be utilized in the investigation.

Quality assurance and quality control (QA/QC) guidelines are set out in federal, state, and at times local agencies' guidance. Laboratory quality assurance objectives have been established to develop and implement procedures for obtaining and evaluating water quality and field data in an accurate, precise, and complete manner. In this way, measurements provide information that is comparable and representative of actual field conditions. Quality control is maintained by requiring the analytical laboratory to perform internal and external QC checks. The laboratory QC includes:

- **Accuracy** — the degree of agreement of a measurement with an accepted reference or true value.
- **Precision** — a measure of agreement among individual measurements under similar conditions. Usually expressed in terms of standard deviation.

- **Completeness** — the amount of valid data obtained from a measurement system compared to the amount that was expected to meet the project data goals.
- **Comparability** — express the confidence with which one data set can be compared to another.
- **Representativeness** — a sample or group of samples that reflect the characteristics of the media at the sampling point. It also includes how well the sampling point represents the actual parameter variations that are under study.

This includes the cleanliness of glassware, purity of solvents, instrument maintenance, and record-keeping. Individual laboratory QC selection criteria may include analysis spike samples, duplicates, and specific standards to insure the accuracy and precision of chemical analyses. The data must be reported accurately, with full laboratory documentation for valid results. Statistical and mathematical procedures for data reduction, instrument calibration, and matrix interferences should be reviewed to monitor performance. These procedures provide checks on cross contamination, but should not be utilized to alter or correct established analytical data. The individual analytical procedure or site sample matrix may cause site-specific deviations that must be taken into account when interpreting data. This approach is used together with the field sampling protocol to assure validity of sample measurement from field to lab.

Units of concentration should be consistent from report to report. Usually analysis results are reported in milligrams per kilogram or liter (ppm). Depending on the laboratory, and the sample matrix (soil or water), analyses may range into the micrograms per liter (ppb) range. It is important to remember that the laboratory will use the method detection limit, but the hydrogeologist should check to be sure that is what the laboratory delivers. While this seems straightforward, many times confusion results if detection limits of the analytical instruments vary, or if the consultant requests a concentration unit change for some reason. The data may be reported as "not detected," which means that it was not detected at the method detection limit for that instrument for that day. The units reported by the laboratory should always be those used and discussed in the investigation narrative and data reporting to avoid confusion.

LONG-TERM TRENDS OF CHEMICAL DATA COLLECTED IN MONITORING WELLS

As the project progresses, a stream of data is collected from the monitoring wells and will form the database of the site. This data should also support a declining trend that will most likely form the supporting rationale for ultimate site closure. The overall trend is anticipated to become lower and ultimately asymptote toward the negotiated cleanup level concentration. This could include an asymptote toward the regional geochemical baseline, or known regional contaminated conditions. The data trend and its quality control become very important as the concentrations decline and approach the cleanup level (see Figure 3).

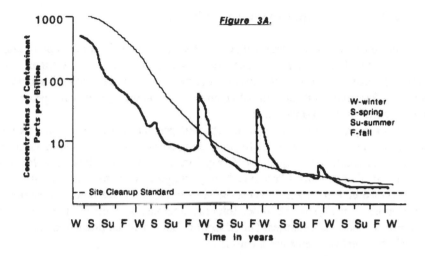

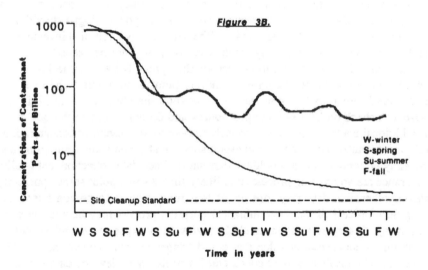

Figure 3 A. Contaminant trends in monitoring wells following source removal. This displays the trend expected for the contaminants to follow once the source is removed and the remaining pollutants become degraded and diluted. Monitoring well data spikes correspond with precipitation where recharge and leaching have remobilized the residual contaminants, causing a brief concentration increase followed by a return to the declining trend. B. Contaminant trends with partial source removal. These trends appear to show some contaminant source removal, but the anticipated diminishing trend is not apparent. This trend may indicate that while the main source was removed, a second source or large residual remains in soil and groundwater and is a continuing source. Contaminant levels fluctuate seasonally but are not declining. More assessment work should be done to ascertain contaminant source and quantity.

Laboratory Data Reporting at the Detection Limit

Since the water quality data will deal with contaminants in ppm or ppb ranges, the accuracy of the analytical instrument during the analysis is critical. The analytical instrument may have varying accuracy on given days and the accuracy near the detection limit, or limit of accurate measurement, may have profound implications in the reported value.

For example, if a contaminant is reported present at the same value as the detection limit, is it really there? Residual concentrations of a contaminant in an aquifer may create long sequences of very-low-concentration "hits." The laboratory report of contaminant "presence" may trigger a regulatory action based on that reported concentration. This can be very important since potential contaminants xylene, methylene chloride, toluene, acetone, and other chemicals are commonly used in laboratories. The laboratory should check procedures to ascertain if analytical problems are a cause.

Extreme care should be used for any data reported at or near the detection limit of an analytical instrument. The data should be discussed with lab personnel through all phases of analysis regarding reported chemical variations and interpretation. When data are sent to the overseeing agency, the agency typically accepts the analytical information as correct, so the hydrogeologist should have confidence that the numbers are "real." Otherwise, costly additional investigation and compliance monitoring may result when they may not be needed.

It is usually a good idea to collect another groundwater sample initially if anomalous analytical results arise. Since one point is not a trend, the resample and analysis are a small cost to check data beyond site-observed trends. It is also wise as a check on field sampler procedures and decontamination thoroughness. If additional contaminants not previously known to be present, or concentrations of contaminants, vary widely from established trends, then some resampling and possibly protocol revisions could be warranted. Since data collection and quality assurance are an ongoing process, it is likely that at some point some apparently incomprehensible data may turn up. If the trend appears to be consistent, the long-term process of that contaminant desorption process may be what you are observing. Data interpretation should be reviewed by experienced analysts before accepting a concentration value that could trigger the enforcement action. The experienced consultant should check and immediately review the data for accuracy. In this way, the best information is acquired for the time and money.

Data Management

Both investigation and periodic monitoring programs generate huge quantities of information. A data management system is needed for each site so that the information is accessible and can be manipulated for needs of reporting and analyses for trends. This should be done at the beginning of site work starting a historical database, and the information can be retrieved when desired. If a long-duration project, then data logged by well may come in at quarterly intervals together with quality control samples for years.

This is a practical problem in data manipulation, and computers and spread-sheet programs are useful for data storage. Computers and database spreadsheets are available depending on the use and type of project. When linked with graphing programs, the data can be reduced to show detected contaminants and time trends. For example, concentration or separate phase product may be plotted against well water levels. Usually a data set showing one well with chemical data by sample date, laboratory sample number, and contaminant may be used in reports or required presentations. Data management methods should be thought out in the initial planning stage if the project anticipates a large data accumulation.

Example: Is a Low Concentration of Contaminant Actually There?

A 1000-gallon gasoline spill from a subsurface storage tank allowed gasoline to migrate through a very porous vadose zone and into an aquifer of sand and gravel. Transmissivity values for the aquifer range from 100,000 to 1,000,000 gallons per day per foot. A regional groundwater pumping field about 1200 feet away pumps over 2 million gallons per day for potable water consumption. The bulk of the spill was cleaned, but some product entered groundwater, and mon-itoring wells were installed to ascertain benzene migration toward pumping wells. A simple benzene linear transport model predicted a possible arrival time to observe whether the plume was in fact moving to the pumping wells.

Three months of weekly groundwater monitoring revealed contaminants were not detected. The contaminant should not have passed the monitoring wells based on the assumed groundwater velocity. The plume was not observed in the mon-itoring well array. Then one sample showed a one-time occurrence of very low concentrations of xylene. According to the benzene transport model assumptions, the xylene could indicate a trailing part of the plume. Benzene, which was assumed to have preceded the plume, was never detected in any sample at any time. Given the one "hot" xylene sample and potential threat to a large water supply, the regulating agency was ready to order immediate extraction-well instal-lation and counterpumping. The cost for one extraction well and initial pumping for this aquifer could exceed $100,000; was extraction really needed, based on the laboratory data?

The well was resampled and the results were similar, but near detection limit. Sample blanks were taken for the sampling equipment and the accompanying travel blank; a duplicate sample was sent to another laboratory. A split sent to a second laboratory showed that contaminants were not detected in either the blank or the second sample. The appearance from the data was that the first laboratory had erred — either by contaminating the sample internally, or their instrument had not been properly cleaned and may have induced contamination in the sample.

Now it is up to the judgment of the professional: if additional xylene "hits" are observed, it would be a plume moving toward the pumping wells. The immediate extraction-well installation did not seem necessary given the previous "not detected" data. The model predicted plume would be preceded by benzene first and xylene last at an assumed velocity. Additional sampling revealed all the fuel constituents for which the program monitored were not detected. One data point does not indicate a trend, and additional monitoring must now be made for confirmation. The low level of xylene observed was attributed to possible lab error, and xylene presence was not observed again over several additional months. The observed data was assumed to be a one-time occurrence. Contaminant plume location needs to be verified by continual monitoring presence in wells. By narrowing the problem to possible laboratory problems, and lack of groundwater contaminant trends, a costly remediation that was not needed was avoided.

The point is that limited data must often be evaluated and reported. One piece of data may initiate a regulatory action. Errors may occur at any point in sampling and analysis, and a resample and laboratory check is necessary. Of course, analytical laboratories should not become scapegoats just because the data does not make sense to the hydrogeologist. The hydrogeologist's approach to the problem may need adjustment. Laboratories may make mistakes or report data at very low concentrations that, based on a one-time occurrence, are extremely difficult to substantiate. Analytical data need to be taken in the context of the spill problem, sampling, analysis, aquifer conditions, and all observed data trends. Taken together, these may signal presence of a contaminant plume, or it could be a "ghost." These data quality problems, given limited information and regulatory implications, can occur but are problems with which the consultant must deal.

CHAPTER **8**

Aquifer Analysis

INTRODUCTION

Aquifer analysis is required for groundwater remediation because the water must be extracted and treated at the surface. Three aspects of aquifer analysis directly relate to remedial action design:

1. the field test to ascertain the optimum yield of the well and allow an estimate of the extent of well influence can be directly observed; from this the hydraulic containment "capture cone" may be estimated;
2. the mathematic analysis of aquifer test data (transmissivity and hydraulic conductivity), helps to estimate the quantity of water that the remediation system must handle;
3. allows sampling of groundwater for contaminant concentrations collected in chemical time-series sampling, and interim water treatment may allow for a field evaluation of a small remedial system operation.

This analysis yields basic design data upon which the project success depends, and in the author's opinion is arguably the most important step where the remedial engineer and hydrogeologist work together. While groundwater extraction and plume containment is costly, installation of subsurface physical structures for plume containment and use in remediation are vastly more expensive. Hydraulic plume containment is accomplished and groundwater must be extracted for treatment anyway, unless the site problem is so huge that physical containment is a viable way to buy time for ultimate site cleanup, which is usually beyond the financial means of the (smaller) client. Thus, the well yield for the required capture is known, and that quantity of water will have to be treated, and this data can then be used for computer simulation of different pumping scenarios.

CONSIDERATIONS PRIOR TO THE AQUIFER TEST

The aquifer test program will typically follow the site definition study. Thus, monitoring wells have been installed at locations at which both chemical and potentiometric data can be collected. This "thinking ahead" allows well positions to anticipate use in the aquifer test. Ideally, while wells are installed for specific head and gradient data, often there must be a compromise for well positioning. However, the data is typically sufficient to ascertain the hydrogeologic and chemical parameters desired.

The proper review and conceptualization of site geology cannot be overemphasized. The site geologic assessment should have yielded most of the basic geologic constraints on the hydrogeology, including but not limited to: aquifer porosity and hydraulic conductivity estimates, predictions of possible boundary conditions, anticipating delayed yield or drainage, secondary recharge or discharge in the aquifer, and aquitard leakage during the test. The site geology should have been defined adequately in the subsurface study, so lithologic cross sections should allow prediction of possible well hydraulic effects. This early conceptualization should help anticipate some well responses or may indicate that additional measurements may be desirable at specific wells.

Potential Boundary Conditions

Boring logs and cross sections should be checked for possible boundary conditions. The lateral and horizontal extent of the strata could indicate possible conductive units where water yield could increase or decrease. If the shape of the cone of depression changes, or measured water levels in observation wells show change or no change, a boundary may have been encountered.

Delayed Drainage

Once the test begins, the strata will yield water but additional water may drain as time passes even though water levels in wells show decline. Water draining from more fine-grained sediments may take longer than from more coarse sediment. Delayed drainage may be recognized on field plots where the drawdown curve steepens, flattens, then steepens again. Often the pump test must be run for a long time to observe this drainage.

Secondary Permeability

Observation wells may show time lags or very rapid water-level changes depending on the hydraulic connection of the geologic materials. This may be observed in fractured rock and depend on presence of fractures, joints, solution cavities, or other pathway. If the wells are not connected then the drawdown data may be difficult to obtain and interpret.

Aquitard Leakage

Recharge to the aquifer may occur through overlying or underlying aquitards. If observation monitoring wells are placed in the aquitard or near the aquitard contact, changes of water levels could indicate leakage. The site investigation data of the aquitards may yield information on the possibility of leakage.

Underground Utilities

The presence of utilities should be known, especially if wells are located close to them. Storm and sewer lines' leakage may affect the response of monitoring wells during the test.

Insufficient or Poor Well Development

The pumping well and observation well construction and development are extremely important. If the wells are not properly developed before a test, typically early data (usually several hours) may not be representative, or appear unexplainable (erratic) due to borehole "skin" effects, and the well may produce turbid water. In many cases, what this early test data represents is well development occurring during the early stages of a test, and discharge may be less than the well can actually produce. Well development should precede a test to enhance data collection, particularly early test data.

INSTANTANEOUS DISCHARGE TEST (SLUG TEST)

Instantaneous well testing or "slug" testing can be used to measure hydraulic conductivity around boreholes, such as production, monitoring to test wells (Bouwer, 1989). The slug test is relatively quick and inexpensive, and can be used to collect data from a site and make estimates of transmissivity from the hydraulic conductivity. Hydraulic conductivity of an aquifer can be estimated from the rate of rise or fall of the water level in a well after a slug of known volume is either instantaneously introduced or removed from the existing water column (see Figure 1). The main advantages of performing slug tests as opposed to pumping tests are:

1. Less expensive to perform.
2. Less equipment needed.
3. Less time to obtain data in the field.
4. Shorter data interpretation/reporting time.
5. Can be used where pumping data may not work (i.e., low-yield conditions).
6. Can be used in small-diameter wells.

Disadvantages include:

1. Transmissivity and hydraulic conductivity values are estimations at best in most cases.
2. Aquifer is not stressed sufficiently to properly evaluate area of influence from tested wells; may test sand pack if unconfined.

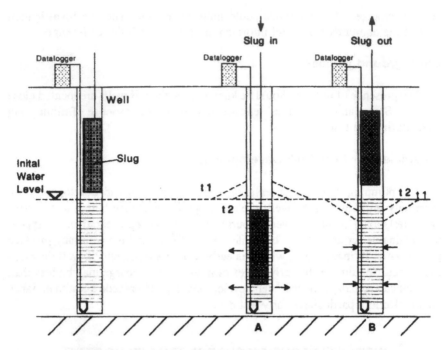

Figure 1 Conceptual slug test procedure. A slug of a known volume is lowered into a well
displacing water (A). The water level is monitored until it returns to the original
static level. The slug is then removed and the test run again as a check (B),
observing the water level rise and return to the original level. A hydraulic conductivity value can be calculated from the data.

3. Only applicable to low-yield aquifers.
4. Not applicable in large-diameter wells
5. Can yield very erroneous data if test well is not properly developed.

Equations developed by Ferris and Knowles (1954), Cooper et al. (1967), and
Bouwer and Rice (1976) are commonly used methods to evaluate slug test data.

These data are sometimes used with computer modeling to estimate pumping
rates of extraction wells for site cleanup. When used in low hydraulic conductivity
aquifers, this information modeling may give useful possible pumping scenarios
and estimated extent of pumping well's influence. However, the author's experience is that this use should be reserved only for very-low-yield aquifers, since
short-term pumping tests may not be feasible to determine long-term pumping
rates. Also, chemical water-quality data cannot be collected from a pumping well,
as would be anticipated in a pump-and-treat scenario. Nonpumping tests do not
stress the aquifer by pumping, so any calculated pumping extraction rate is an
approximation that may or may not be real. A computer model would estimate
well yields, but actual remedial pumping system performance is the measure of
success, so field pumping data is invaluable. This means the observed capture
extent of contaminated water and extracted contaminant concentration data
greatly assist remedial equipment design.

PUMPING TESTS

A pumping test will stress the aquifer and allow water-level changes to be observed. The purpose of a step-drawdown/well recovery test is to (1) estimate aquifer transmissivity, (2) select the optimum long-term discharge rate for a constant-rate discharge test, and (3) identify wells in hydraulic communication with the pumping well for monitoring during the second part of the test, the constant-rate discharge test. A step-test can be completed within a relatively short timespan (usually 6 to 10 hours), and only a single well is needed for the estimation of transmissivity. In comparison, a constant-rate discharge test typically requires a minimum of 24 hours of pumping from the pumping well and at least one observation well within the anticipated radius of influence to calculate transmissivity and storativity. If the saturated aquifer thickness (t_o) is known, hydraulic conductivity can be estimated. The step-drawdown/well recovery test consists of two phases; the step-drawdown test and recovery residual test.

STEP-DRAWDOWN TEST

The step-drawdown test involves pumping a well at an initial discharge rate (Q_1), which is incrementally increased (stepped) while drawdown (s) is measured in the test well at various time intervals. The purpose of the test is to ascertain the optimum yield at which the constant duration test should be performed. Ideally, the step-drawdown test utilizes three or more different discharge rates or steps $(Q_1, Q_2, ...)$ as a minimum, with each subsequent flow rate increased from the previous flow rate. Under most conditions, rate increases will vary. Therefore, particular attention to subsurface geologic conditions, well design, and available water column in the well are critical to estimate potential flow rates before a test is started. An initially "conservative" flow rate is recommended to evaluate aquifer response and material capability (see Figure 2).

The duration of a particular step depends on test field observations. The observed water level with respect to available water column in the pumping well dictates the length of a step and potentially how many steps a test will have. The target duration of each step should be at least 30–60 minutes, so that enough data points can be plotted to establish a trend on the semi-log graph. Sixty minutes is preferable unless drawdown conditions (i.e., dewatering) preclude a step from running that long. If the water level within a step does continue to decline, you may be looking at the last step in the test. If this occurs in the first step of a test, the pump may be shut off, allowing full well recovery and restarting the test at a lower pumping rate. Or, it may mean that a step test cannot be performed properly due to low-yield conditions, and slug tests may be a viable alternative. Likewise, if very little drawdown occurs in a step and no changes in drawdown occur over a 30-minute period, proceeding to the next step is usually justifiable.

The selection of the discharge rates for each step must consider the water level in the discharge well observed during the previous step. It is better to select a conservative rate increase than to overstress the aquifer and dewater the well. The

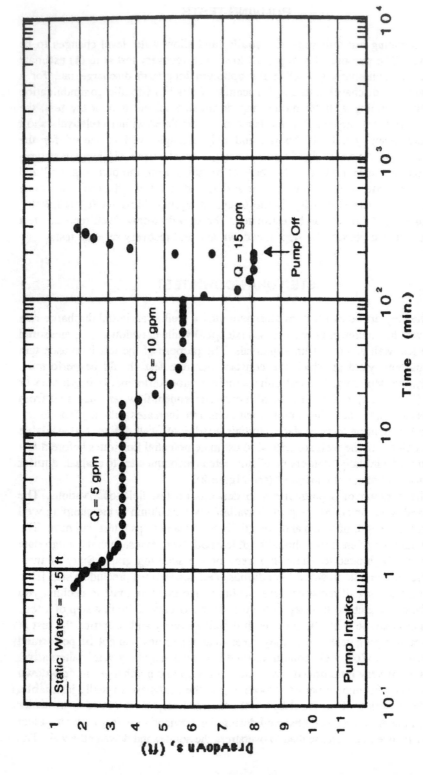

Figure 2   Step test. The step test allows the pumping well to be pumped at increasing rates and for the drawdon to be observed. As each step pumps more, the well draws down to some depth where the drawdown becomes constant. The pump intake depth is noted so that drawdown below the intake does not burn the pump. The step test here indicates that a 15 gpm yield should be sustained. Once the pump is truned off, recovery measurements are collected.

effects of pump rate increases for different steps can be tracked by constructing a simple graph before a step-drawdown test is conducted. The graph should include: the available water column present in the well, well design, aquifer/aquitard or aquiclude relationships, and location and depth of the submersible test well pump intake. Some suggested rate criteria are listed below (Palmer et al., 1992).

1. If drawdown levels out at less than 25% of available water column, flow rate can be increased.
2. If drawdown levels off between 25 and 50% of available water column, flow rate may be increased. One should use field judgment on expected drawdown based on observed drawdown from previous step.
3. If drawdown levels off between 50 and 75% of available water column, flow rate may or may not be increased. Again, look at previous data and perhaps select a conservative increase in order to prevent dewatering.
4. If drawdown is greater than 75% of available water column, a flow rate increase could dewater well. A conservative pump rate increase may be possible but should only be attempted if previous data indicate such an increase might be possible.
5. If drawdown is greater than 90% of water available water column and dropping, prepare for recovery test. Shut off the pump and begin the recovery measurements.

The above criteria are only guidelines. Obviously, each test site will vary according to hydrogeological conditions; aquifer/aquitard relationships, well design, and the amount of water available in an aquifer. One should increase pumping steps "conservatively" if you are not sure how the pumping well will respond to an increase in the discharge rate. Additional steps are preferable as opposed to insufficient steps to evaluate potential well yield and estimate transmissivity.

PUMP TEST EQUIPMENT

Proper equipment is necessary to conduct a pump test. Equipment depends on the types of tests to be performed, site logistics, well design, presence and type of contamination, and number and location of observation wells. Because the field team may be at the site for several days, it is wise to be sure the required equipment is on hand rather than running about trying to find or buy it. The following suggested equipment list is typical but can vary depending on test requirements (Palmer et al., 1992):

- Submersible pump (stainless steel components preferred; equipped with a check-valve, and appropriate control box)
- Pump discharge line or piping (compatible to discharge pump)
- Pipe dog or tripod (to secure pump in well)
- Flow-valve control (gate or ball-valve) and flowmeter system
- Discharge piping or line (flat-lying flex hose is good)
- Power source (AC power or suitable portable generator and fuel)
- Stop watches (two as a minimum)

- • Water-level measuring devices (electric sounder, steel tape, oil/water interface probes)
- • Data-logger and pressure transducer
- • Semi-log and log-log graph paper
- • Rulers, French curves, mechanical pencils
- • Scientific calculator
- • Site safety equipment as outlined in a site-specific safety plan.
- • Data sheets to record drawdown, recovery data, time, well numbers, pumping rates, etc.)
- • Tools, fittings, etc. as required
- • Lighting (flashlights and/or lanterns)

In addition to field equipment, the hydrogeologist, geologist, engineer, or technician should have the following documentation in the field during a test:

1. Boring logs for wells to be pumped and monitored.
2. Well-completion details of wells to be pumped and monitored.
3. Well-development data sheets (if available).
4. Groundwater sampling purge data.
5. Latest groundwater chemical data collected from monitoring network.

Pretest Field Procedures

Obtain a topographic or facility map and check access constraints, security, potential traffic problems, and integrity of test wells. Identify potential or required discharge points and distances from test wells. Secure all required permits, rights-of-way, variances, etc., from proper state or local agencies, or private parties. Review lithologic logs and well-construction details. Obtain and review any pump histories for the test wells. For example, if the wells were developed, review the development discharge rate. Also, review sampling records if available since purging data can be very useful to estimate well yield(s). Assemble all necessary equipment to perform the pump test. Make sure that all equipment is functional before you go to the field. Decontamination of equipment at the start, and between each well use, is essential on all contaminated sites. Since this is a costly portion of the field work, preparation is key.

Pretest Well Water Level Monitoring

Selected pumping wells and "background" or observation wells should be monitored for at least one day (24 hours) prior to a test. The objective of pretest well monitoring is to identify changes in water levels due to diurnal changes such as tides, effects of irrigation, or pumping from local domestic or municipal wells. Graphic plots of water-level changes vs. time are very useful in identifying water-level fluctuations unrelated to the pump test. Water levels can be periodically measured by hand, or by using a transducer/datalogger system to record data prior to a test.

Suggested Field Procedures

1. Measure static water level in the discharge well. Measure depth to bottom of well casing, and record depth to accumulated fines or possible well obstructions. Calculate available water column.
2. Install submersible pump in the well to the desired depth of placement. Usually the pump intake is placed opposite the bottom of the screened interval (placement will vary according to well construction and geologic conditions). It is a good idea not to set the pump on accumulated silts or clays in the bottom of the well because of the potential for pump damage. Ideally, test wells should be properly developed before a pump test.
3. Observation wells should be identified that may be influenced during the test, and water levels in those observation wells should be measured and recorded. Water levels can be measured in the observation wells periodically to evaluate radius of influence and development of the cone of depression produced by pumping. Remember to compare data with background data so changes in observation wells are related to pumping and not caused by natural phenomena. Decontaminate sounders between each well to prevent well cross contamination.
4. Pumping discharge rates be maintained within a 10% range. Check discharge rate frequently to be sure that the pumping rate does not fluctuate. Identify the reference point from which water levels are measured (i.e., top of well casing, top of well box, etc.). Field data plots are very useful during a test and assist in determining when a test should be terminated (if rapid dewatering of the well happens) or if the electronic datalogger fails (see Figure 3). Do not change water-level measuring device during a test, unless required due to equipment failure. Always have a backup measuring device.
5. The times the pump was started and shut off should be noted. A check valve should be installed to prevent water from running back down into the well and recharging it "artificially." If the test well or wells are part of a monitoring network and have historically contained chemicals, decontaminate equipment before every test, between tests, and after the last test. Field measurements to be taken during the test include the following: time since pumping started; time since pumping stopped (recovery); depth to water; discharge rate. The greatest fluctuation in water levels typically occurs at the test start, and occasionally at the end of a test. Therefore, water-level measurements at the beginning of a test should be recorded more frequently. Ideally, a datalogger/transducer system should be used to collect drawdown data. The following time intervals are guidelines; readings may vary according to site conditions.

Suggested Measurement Intervals (Driscoll, 1986)

Elapsed time	Water-level measurements
0–10 minutes	Every 0.5–1 minute
10–15 minutes	Every minute
15–60 minutes	Every 5 minutes
60–300 minutes	Every 30 minutes
300–1440 minutes	Every 60 minutes
1440–end of test	Every 8 hours

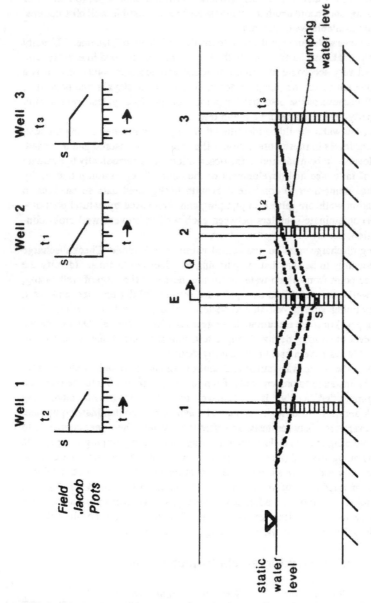

Figure 3 Observations during constant discharge test. As the cone of influence of the pumping well E arrives at the observation wells at different times, the drawdown s and time t are plotted as Jacobs plots. Changes in drawdown may indicate boundary conditions. When well response field plots are compared to site boring logs, it may reveal other geologic aspects not observed directly in boreholes.

Constant Discharge Test

The constant discharge test is started once the test well has fully recovered from the step-discharge test. The constant test would be run at the optimum well yield determined in the step test, and run for some longer period of time (typically 24 hours, or longer), and followed by the recovery observations. The data collection procedures for the constant test are essentially the same as the step-test discussed above. The constant discharge test allows for long-term observation of the cone of influence and chemical sampling during the test. Occasionally, some problems may arise during the test, such as a boundary condition (where the pumping well water level declines) to the point where the well cannot pump. In this instance, the recovery test would be started immediately as described above. Delayed drainage may occur where water levels may decline, then increase in observation wells; hence, field plotting data can help determine the time and rough location of the recharge. There can even be contaminant invasion into the pumping well, such as when a floater drains in the vicinity of the extraction well and is pumped to the surface. This is a pollutant containment problem for which the field crew must be prepared if site conditions indicate the possibility.

Well Recovery Test Field Procedures

Measurements of water levels begin immediately after the pump has been shut off. Measurements should be made at the following time intervals as a minimum:

Well Recovery Measurement Intervals

Elapsed Time	Water-Level Measurements
0–5 minutes	Every 0.5 minute
5–10 minutes	Every minute
10–30 minutes	Every 5 minutes
30–end of recovery	Every 10 minutes

Water-level measurements can be terminated if less than 0.01 foot of change in water level is measured over a 30–60-minute period. Unchanged readings for this period of time typically indicate that recovery to original static-water level will take a long time. A pump test may be considered done if recovery occurs to within 80–90% of static-water level.

A water-level return to a depth shallower than (above) original static level as measured from ground surface may be observed. This may be the result of normal diurnal changes or a "rebounding" effect due to pumping. Often, a water level will rise slightly above the original static level, then fall again close to the original static level. Plot recovery measurements (drawdown vs. time) during well recovery. The pump or any down-well test equipment (e.g., transducers) should not be removed from a well until the recovery test has been completed. Premature removal of pump and discharge line equipment will give erroneous water-level data.

Residual Recovery Test

The residual or recovery test involves monitoring well recovery after the pump has been shut off. Recovering water-level measurements are collected until original static water level has been reached. If a test is being performed in an area where low-yield aquifers are being tested, well recovery may take 24 hours or longer to fully recover. In a low-yield aquifer, 80–90% recovery may be acceptable. The last 10–20% of recovery may require some time to collect, and may not be cost-effective to collect entirely.

REVIEW OF SELECTED PUMPING TEST DATA ANALYSIS

The following is a brief review of data collected from the constant discharge test. Although this review is presented, it is not a substitute for completing rigorous course work in a groundwater hydraulics class. Only two methods of analysis are presented as examples and all cases of aquifer analysis are not reviewed. The reader is referred to the appropriate university classes and texts on groundwater data analysis (for example, Driscoll, 1986; Heath, 1982; Theis, 1935; Walton, 1962, 1970) (for leaky aquifers and partial penetration, see Hantush, 1956, 1960, and 1962; for unconfined aquifers, see Neuman, 1972, 1975).

Theis Method

Theis (1935) noted that during a pump test in a well that penetrated an extensive confined aquifer, if the discharge rate is held constant, the area of influence increases with time. Theis was able to conceptualize the formation of the cone of depression in this type of subsurface environment and realized that the rate of water-level decline when multiplied by the storativity and summed over the area of influence, equaled the discharge rate (Kruseman and DeRidder, 1990; see Figure 4). Theis also believed that as long as water was continually being removed from storage in an isotropic aquifer, drawdown would continue over time and that theoretically, no steady-state flow would exist. Theis' non-steady-state equation is as follows:

$$T = \frac{114.6 \, Q \, W(u) \text{ in days}}{s} \quad \text{and} \quad S = \frac{Tut}{1.87 \, r^2}$$

where:
- T = transmissivity of aquifer in gpd/ft.
- S = storativity of aquifer (dimensionless)
- Q = pumping discharge rate (gpm)
- s = drawdown in cone of depression in feet
- t = time since pumping started, in days
- r = distance to pumping well center in feet
- $W(u)$ = "well function of u"; an exponential integral (infinite series).

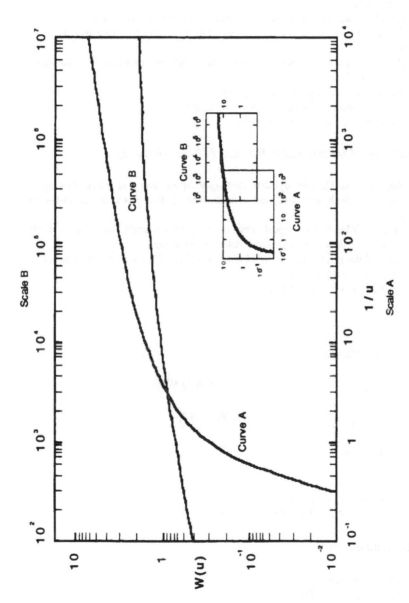

Figure 4 Theis Curve of dimensionless drawdown W(u) to dimensionless time (1/u) for constant discharge from an artesian well. (Modified from Reed, 1980, in, U.S. Department of Interior, 1985.)

The Theis equation follows Dupuit's Assumptions, which are:

1. Aquifer is seemingly infinite in areal extent.
2. Aquifer is homogeneous, isotropic, and uniformly the same thickness and horizontal.
3. Prior to pumping, potentiometric surface is horizontal (i.e., flat).
4. Aquifer is pumped at a constant discharge rate. The pumping well is 100% efficient.
5. Pumping well fully penetrates aquifer, thereby receives water from the entire thickness of the aquifer.
6. No induced recharge during the pump test.
7. Water is discharged instantaneously with drawdown.
8. Potentiometric surface has no slope.

Type Curve of Match Point Method (Theis Method)

1. Take W(u) vs. (1/u) or (u) curve and superimpose over s vs. t field data curve.
2. Make sure both horizontal and vertical axes of data plot and type curve are parallel.
3. Select a "Match Point" (preferably on an even log scale; 10, 100, or 1000). Avoid early data matching. Later data are preferable.
4. Match Point gives you four coordinates: (1) W(u), (2) 1/u or u, (3) s (feet), (4) t (min., converted to days).
5. Calculate lower integral equation:

$$u = 1.87 \ r^2 \ S/Tt$$

To get u; take 1/u and use (u).

$$\text{or:} \quad T = 114.6 \ Q \ W(u)/s$$

$$\text{and:} \quad S = Tut/1.87r^2$$

To solve Equation 1:

a. Get Q from test.
b. Get s from Match Point.
c. Get u from Match Point (reciprocal of 1/u).

To solve Equation 2:

a. Get T from Equation 1.
b. Get u from Match Point.
c. Get t from Match Point.
d. Get r from field measurements (distance between pumping well and observation well in feet).

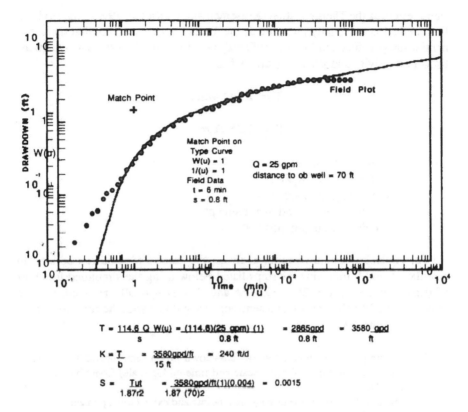

$$T = \frac{114.6\ Q\ W(u)}{s} = \frac{(114.6)(25\ gpm)\ (1)}{0.8\ ft} = \frac{2865gpd}{0.8\ ft} = \frac{3580\ gpd}{ft}$$

$$K = \frac{T}{b} = \frac{3580gpd/ft}{15\ ft} = 240\ ft/d$$

$$S = \frac{Tut}{1.87r2} = \frac{3580gpd/ft(1)(0.004)}{1.87\ (70)2} = 0.0015$$

Figure 5 Example of Theis analysis (see text). The field data is plotted and overlain on the Theis type curve and the data read off at the Match Point, convenient at 1/u = 1 and W(u) = 1. Note that at the end of the test, a possible recharge boundary was observed, since the curve flattens — a valuable piece of data that was observed because a long-duration discharge test was conducted.

Note: (1) t is always expressed in days; (2) data are plotted on same scale graph paper as the type curve (see Figure 5).

In addition to plotting pumping data, recovery data (pumping well and observation wells) or what is known as "residual drawdown" can be used to calculate transmissivity and storativity. Residual drawdown vs. time since pumping stopped is plotted on log-log graph paper. The Theis curve is used to find the match point and T- and S-values are found using the two Theis equations. Storativity can only be calculated if data from an observation well is used.

Cooper-Jacob Graphical Method

The Cooper-Jacob Graphical Method (Cooper and Jacob, 1946) for evaluating transmissivity (T) is based on the Theis equation, although the conditions for its application are more restrictive. For values of u ≤0.05 (usually u is greater than 0.05 early in a pump test), the modified nonequilibrium equation will give the same results as the Theis equation (Kruseman and DeRidder, 1990). The straight-line method is essentially the same as the Theis equation, except that the exponential integral function W(u) is replaced by a logarithmic term. The equations for transmissivity and storativity are as follows:

$$T = 2.3 \ Q/4\pi\Delta s$$

$$S = 2.25 \ Tt_o/r^2$$

where: T = transmissivity of aquifer
  S = storativity of aquifer
  Q = pumping discharge rate
  t_o = time at zero-drawdown intercept
  r = radius of the pumped well.

For nonleaky, confined aquifer analysis, the Jacob's Method can be used to calculate T and S following the steps listed below using the simplified formulas of (time-drawdown) $T = 264 \cdot Q + \Delta s$ and $S = Tt + 4790 \cdot r^2$ where t is the intersection of the time-drawdown semi-log straight line with the zero-drawdown axis (see Figure 6A):

1. Plot pump test data time (t) versus drawdown (s) on semi-logarithmic graph paper (drawdown on arithmetic scale and time on log scale) from observation well data.
2. Draw best-fit straight-line through data points and extend line up to zero drawdown (s=0).
3. Measure slope of the straight-line (Ds for one log cycle — slope is defined simply as "rise over run").
4. Read value of t_o at intersection of zero-drawdown line. Value read will be in minutes; convert to days to estimate storativity.
5. Solve for Equation 1 to get transmissivity (T).
6. Use observation well data (drawdown and distance to pumped well) to calculate storativity (S). Use T-value from Equation 1.

The Cooper-Jacob data plot should be used during every pump test to track the progress of the pumping well and observation wells. Deflection of the plotted data may indicate boundary conditions of other local effects. Typically, drawdown and time data points are taken from a datalogger (pumping well and close proximity observation wells), and plotted on semi-log graph paper. Later in the test, a best-fit straight line can be used to estimate transmissivity and used with observation well data to estimate storativity of the aquifer.

Distance-Drawdown Method

When several observation wells are monitored during a constant-rate discharge pump test, the distance-drawdown method of aquifer analysis can be used to calculate transmissivity and storativity. Three observation wells are required as a minimum to use this method of analysis. Using drawdown data from three observation wells recorded at the same time, these values can be plotted on semi-log graph paper; drawdown in feet on the arithmetic axis and distance from the pumping well in feet on the log-axis (see Figure 6B). The relationship of draw-down to distance from the pumping well results in a straight line when plotted on semi-log paper. The modified nonequilibrium equation can be transformed to use distance-drawdown plots to calculate (T) and (S):

$$T = 528 \ Q/\Delta s \qquad \text{and} \qquad S = 0.3 \ Tt/r_o^2$$

$$(\text{or rewritten, } S = Tt \ / \ 4790 \ r^2)$$

where: T = transmissivity of aquifer in gpd/ft.
  S = storativity of aquifer (dimensionless)
  Q = pumping discharge rate (gpm)
  t = time since pumping began (in days)
  r_o = zero drawdown intercept of extended straight line through a minimum of three data points (in feet).

The intercept at the zero drawdown line is the r_o value at the intersection of the semi-log straight line with the zero drawdown axis. The points on the semi-log plot are connected and the line is extended to the zero drawdown line, similarly to the best-fit line used for the Cooper-Jacob Method. As previously described pump test methods, Dupuit's assumptions are believed to be true for this analysis method. In addition to being able to calculate (T) and (S) using the distance-drawdown method, well-efficiency can also be estimated (Driscoll, 1986). The well efficiency values (percents) obtained from this method are estimations at best. Factors that affect this calculation are well design, subsurface geology, pump test parameters, and well development.

Although many regulatory agencies are interested in the data analyses of computer models, supplementing these pump test data with hand-calculated data is important. By performing hand calculations, you are demonstrating to regulators and potential peer-reviewers that you understand the mechanics involved in quantifying aquifer properties. Pump test data are used to select, develop, and implement remediation. While every computer model can generate T, S, and hydraulic conductivity values, the hydrogeologist alone must determine whether data are representative of site-specific conditions. Hand-calculated data should be used to support or refute computer-model calculations, especially assumptions of model input. It is advisable that both hand-calculated and computer-generated aquifer parameters be compared with collected field data to verify they make sense for the "system" in which you work.

The aforementioned methods are in common use but do not cover all aquifer situations (such as delayed yield correction, leaky confined, etc.), of which an exhaustive review is beyond the book scope. The hydrogeologist running the test program and data analysis is responsible for recognizing the need for additional and appropriate analysis methods given the site-specific aquifer case. The reader

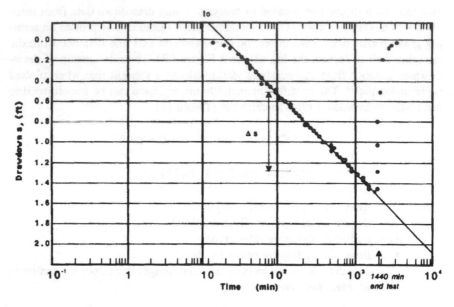

Δ s = 1.3 - 0.5 = 0.8 ft; r = 25 ft (distance to ob. well); Well discharge = 15 gpm

$$T = \frac{264\,Q}{\Delta s} \qquad\qquad S = \frac{Tt_o}{4790r^2}$$

$$T = \frac{264\,(15)gpm}{0.8\,ft} \qquad S = \frac{4950gpm/ft\,(20\,min)}{4790\,(25\,ft)^2}$$

$$T = 4950\ gpm/ft \qquad\qquad S = 0.003$$
$$or \quad 4,560,000\ gpd/ft$$

Figure 6A   Cooper-Jacob method, time — drawdown plot.

is referred to the hydrogeology texts and aquifer analysis methods for further reading (see References).

AQUIFER TESTING IN FRACTURED ROCK

Aquifer testing and analysis in fractured rock differs from that described above. The aquifers are fractured rock covered with alluvial cover and saprolite, or perhaps interconnected with other water-bearing fractures. Recall that the water movement in these terrains is related to the water-bearing qualities of the overlying material and the amount, density, and interconnection of fractures and bedding planes (or solution cavities) in the rock. Typically the fractures close with depth, and wells tend to not produce much water below about 300 feet as a general rule (unless there are additional water-bearing strata, leakage, presence of other water-bearing formations, fractures zones or faults, etc.)

Groundwater flow in fractured rocks is from radial and linear flow. The degree of water movement is related to the interconnection of fractures and the quantity of

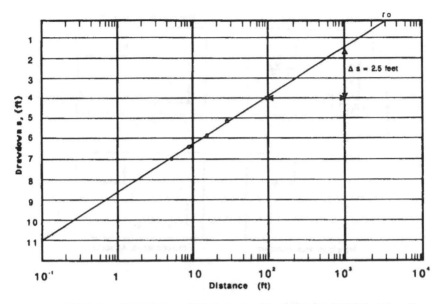

Δ s = 2.5 ft; t = 500 min. into test; Well discharge = 15 gpm, water levels from five site wells.

$$T = \frac{528 \ Q}{\Delta s} \qquad\qquad S = \frac{Tt}{4790 \ r_o^2}$$

$$T = \frac{528 \ (15 \text{gpm})}{2.5 \text{ ft}} \qquad S = \frac{4,600,000 \ \text{gpd/ft}(500)}{4790 \ (3000)^2}$$

$$T = 3,200 \text{ gpm/ft} \qquad S = 0.005$$
$$\text{or } 4,600,000 \text{ gpd/ft}$$

Figure 6B   Cooper-Jacob method, distance — drawdown plot.

storage in the material overlying the fractures in the system as a whole. The testing phase would be similar (i. e., a pumping well and observation wells) but data analysis may need to be performed in the pumping well and observed to see if nearby wells are affected by the pumping. The pumping test and analysis should be factored into the groundwater flow regime model developed for the particular site.

For example, when the extraction well is pumping, observation wells must be observed to ascertain the degree of influence, from which a supposition of fracture interconnection can be made. The extraction well may draw water from the overlying alluvium or saprolite for some time, so "drawdown" may not be readily observed. When the overlying material is drained, then water levels in the pumping well may decrease rapidly to stabilize later, which could represent the water withdrawal from fractures. Other effects could be a directional preference to the cone of influence, elongating in one direction (see Figure 7). Observing recovery may assist in the analysis and trying to ascertain fracture interconnection. Analytical techniques are available for pump test data in fractured rocks (see Jenkins and Prentice, 1982; Maslia and Randolph, 1990; Sen, 1986). Schmelling and Ross (1989) have compiled a list of possible model and analysis approaches together with a summary of data collection techniques in fractured rock.

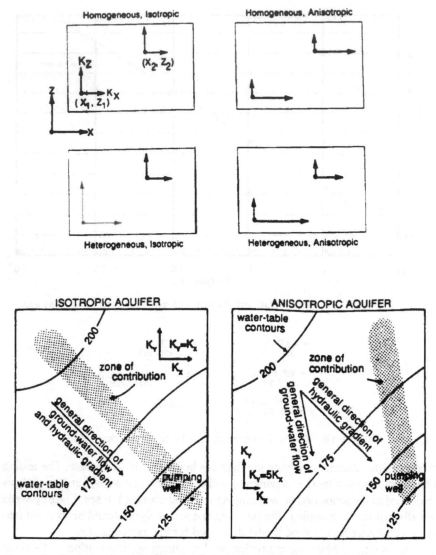

Figure 7 Homogeneity and heterogeneity in aquifers and effects of anisotropy on ground-water flow. (EPA, 1994)

COMPUTER MODELING FOR GROUNDWATER PUMPING AND REMEDIATION SCENARIOS

Computer modeling can be useful in the analysis of groundwater flow in that long-term simulations of different pumping scenarios can be tried, and the effects on the capture cone estimated. Several different types of models are commercially available depending on the proposed use — 2-d or 3-d models, etc. — and can be used on personal computers with relatively little training. A review of computer models is beyond the scope of this book; however, one bias is presented. A computer model is only as good as the data put into it, and that data comes directly from the completeness and accuracy of the site geologic investigation and hydrogeologic characterization.

Computer flow models should be used to estimate physical flow criteria (that is, estimated extent of well influence, etc.) collected from the field. Groundwater flow will be dictated by the geologic framework of the aquifers, and that context should always be kept in mind. When used for remediation, with good field pump test data, flow modeling can be very useful, especially in terms of trying various pumping and contaminant capture schemes (see Keely, 1982, 1984, 1989). In other words, you have to use real-world data, otherwise the model results are purely speculative. If numbers are only assumed to be correct, then the computer simulation may be without any relation to the site conditions (garbage in, garbage out; for an overview of some aspects of computer modeling see Hatheway, 1994).

If a shortcut is to use assumed data (say taking published values and assuming they fit the site based on limited site study) instead of running a site-specific pump test, then will the remediation pump scenario really work? And will it contain the plume and hydraulically control it? Or does the consultant only assume it will work? Does the cost saving in modeling, rather than field testing, actually pay back to the client? This kind of approach could be a great disservice to the client since it can waste his money and not solve his problem. When site-specific, real data is used in the model, then the model simulation can project what could happen long term, or it could be varied to account for lower well yield, or add more wells to get an estimate of conc interference before installing them, and so on. This seems the more proper use of computer models, where the greatest benefit for the client's money and interest is achieved.

Pumping tests are the only tests in which the aquifer is stressed so site-specific pumping effects are observed and measured. These tests may also reveal stratigraphic effects and boundaries, or areas of recharge or discharge that would not be apparent by doing only slug tests. Since any pumping remediation scheme is dependent on pumping, well efficiency, and cone of influence to capture contaminants, pumping tests must be done. If a pumping test is not done, and the remediation depends on a calculated pump rate, neither the capture radius observation of the well nor the true yield of the well is known. Consequently, the design and cost of the remediation are similarly compromised.

POSSIBLE PUMPING APPROACHES FOR CONTAMINANT CAPTURE

The goal of the pumping system is, of course, to capture water and deliver it to the surface treatment system. Depending on the site and problem, many variations of groundwater pumping could be possible. Many variations could be used; however, two basic pumping approaches are presented: pumping from wells (hydraulic conductive conditions, with or without injection wells) and pumping from trenches (low hydraulic conductivity such as silts, clays, and shallow groundwater occurrence).

Pumping from wells is used when the extent of drawdown and yield are sufficient to "capture" the plume. One or more wells could be used with drawdown cones "interfering" with each other to broaden the influence (Keely, 1984). This approach could be used in predominantly sandy or gravelly aquifers, or aquifers with sufficient yield to attain this goal. This approach lends itself to computer modeling since the information collected is usually readily usable in the system. The remedial system can be modeled by adding wells, changing the pumping rates etc. and modeling simulations into the future to estimate system effectiveness (Keely, 1989). The extent of capture can be calculated and flownets prepared (Figures 8 and 9).

Injection wells may be used in various configurations to create mounds or to move or "flush" contaminants toward the extraction well. In this case, the injected water must be captured by the extraction system if flushing is the objective. Otherwise, the contaminated water may move beyond the influence of the extraction well, creating another problem. If clean water is injected to create a mound or barrier to contaminant migration (as toward a potable municipal water well), then the mound would have to be maintained to prevent contaminants from "breaking through." Observation, by pump testing, is very important to ascertain the "capture cones" so that the system operates as the designer has envisioned. Once testing is done, modeling can be very useful in trying different injection/withdrawal configurations, and modeling effects into the future. Water injection in wells, as well as the addition of materials into wells, (i.e., microbes) may only be done with regulatory approval.

When aquifers are very clayey or very low yield, and where groundwater is near to the surface (within about 20 feet), pumping from trenches may be more applicable and cost-effective. The problem of pumping from very-low-yield aquifers is getting yield and extent of influence estimates from a single pumping point. Pumping from wells may not create the desired radius of influence, and the doubling of the diameter does not double the radius of well influence, but rather creates a "sump" that may be periodically pumped (Anderson, 1993; Driscoll, 1986). Consequently, the trench pumping system is essentially using a "French drain" in reverse. Instead of draining water, the trench captures groundwater along the length of the trench by pumping at wells placed at various locations (usually either center or ends). A broad "capture cone" develops as the trench slowly collects water until an equilibrium is attained. Pumping analysis effectiveness is partially mathematic and partially observation to determine the best optimum configuration. In this instance, slug tests may provide some useful data.

SUMMARY

Pumping tests should be done to ascertain the aquifer hydraulic characteristics, and to observe actual pumping conditions prior to installing a remedial system. Any pumping approach should be observed through time to see if

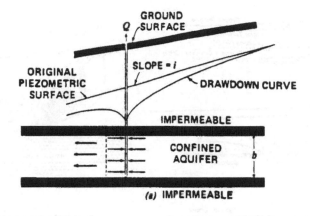

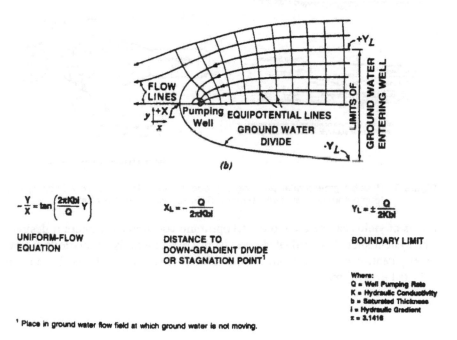

UNIFORM-FLOW
EQUATION

DISTANCE TO
DOWN-GRADIENT DIVIDE
OR STAGNATION POINT[1]

BOUNDARY LIMIT

Where:
Q = Well Pumping Rate
K = Hydraulic Conductivity
b = Saturated Thickness
i = Hydraulic Gradient
π = 3.1416

[1] Place in ground water flow field at which ground water is not moving.

Figure 8 Calculation of groundwater flowlines to a pumping well and estimation of the
groundwater divide created by pumping. (EPA Wellhead Protection, 1993.)

recharging or bounding effects occur. The aquifer test may yield differing data
depending on the time of year, discharge or recharge condition of the aquifer,
nearby regional pumping wells, and the limited duration of the initial pumping
tests. The configuration of the "contaminant capture cone" could change as
aquifer conditions change, which could affect the remediation. Once actual
pumping data are collected, then computer simulation of differing pumping

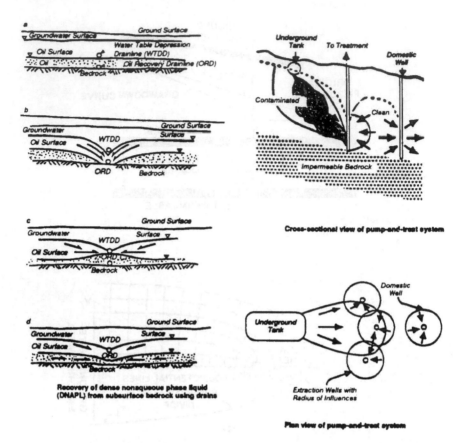

Figure 9   Possible groundwater pumping approaches for DNAPL recovery (a,b,c,d) and
"pump-and-treat" systems. (From U.S. EPA, 1991. With permission)

arrays of yields can be used to model optimum contaminant capture if desired.
The system should be evaluated periodically, with monthly or quarterly water-
level potentiometric maps, and overall effectiveness should be evaluated and
adjusted every 6 months.

CHAPTER 9

Remediation and Cleanup

INTRODUCTION

The goal of site remediation is to restore soil and groundwater quality to pre-contamination conditions. While the ideal goal is to restore the site to "natural" conditions in which all contamination is removed, remediation generally means reducing contamination to the site cleanup goals. In most cases, soil and groundwater cleanup objectives are directed toward an aquifer nondegradation goal, where the groundwater is not degraded and beneficial use is preserved. The intent of government regulations is to restore the site to a preexisting condition; however, this goal is almost never realized since it is impossible to remove every molecule of contamination. Development of regulations and policy for cleanup has become somewhat standardized with a stepwise process for cases to follow (see Figure 1).

The lead agency (local, state, or federal) overseeing the site will usually establish cleanup standards or groundwater protection standards, following a review of the investigation data and regulatory negotiation. These standards are based on numerous criteria, including the site information, contamination type and extent, potential threat to human health, and protection of future soil and groundwater quality. Agencies will usually lean to conservative cleanup standards, although they are usually not set to the most strict standards (as for example the Recommended Maximum Contaminant Levels). The cleanup standard often considers the allowable contaminant concentration left on site, which would not adversely affect human health, soil, or groundwater quality.

The most frequently asked cleanup question is, "How clean is clean?" In other words, what is the appropriate residual contaminant concentration to be left in soil or groundwater, and the degree of effort needed to clean the site while safeguarding health and environmental concerns. As chemical analytical techniques become more sensitive, lower concentrations of contaminants are measured. Remediation goals for any cleanup may desire a return to "natural conditions," but is that a realistically achievable goal? Preexisting natural conditions may result in groundwater naturally contaminated by biologic organic or inor-

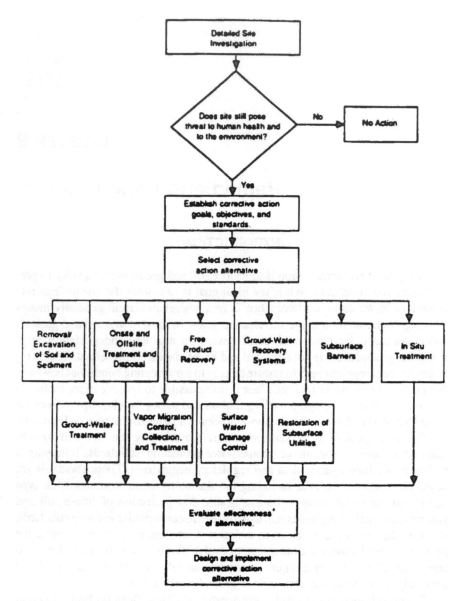

Figure 1   Flow chart of corrective action for undergound tank leaks. (From U.S. EPA, 1985b.
With permission.)

ganics. Residual contaminants trapped in the vadose zone, clayey soil, or aquifer
porosity may be extremely difficult or technically infeasible to remove. If cleanup
standards are enforced at laboratory detection standards, are they achievable and
realistic to protect soil and groundwater? This may be necessary for some chemicals but not others, depending on the health threat. The costs of these decisions
are factored into the site remediation since the responsible party will have to

perform clean up. The site-specific cleanup standard will determine the level of effort and budget required for remediation.

Clearly site remediations may bankrupt responsible parties, and subsequently the government or other responsible parties may complete the task. While the responsible party should be accountable for the problem, the financial resources of governments and industry are not limitless. Cleanup costs must be paid by someone or some group, and the costs and number of cleanup sites continues to increase. Assigning blame in litigation, and determining who pays, is complex, and time-consuming litigation may occur before a group accepts multimillion-dollar liabilities. Management and allotment of funds must be considered carefully or the cleanup could stall and languish, defeating the purpose of the remediation. Delays with regulator negotiation or litigation may delay cleanup, allowing the plume to move and the problem to expand.

Site remediation is a compromise using reasonable judgment based on the available information, applicable regulations, relative public health risk, ability of technology, and cost-effectiveness of the cleanup. Sites will rarely, if ever, be cleaned to preexisting or "natural background conditions." Inflexible positions taken by government, potentially responsible parties, and environmental lobbying groups often delay cleanup efforts. Although the intentions may be good, the delays allow the contaminants to migrate and expand the potential cleanup area while some interested parties cling to unreasonable or utopian positions.

CONCEPTUAL APPROACH TO SITE CLEANUP

Site remediation considers the limits of technical expertise, existing government regulations, environmental protection desired, and economics. Costs for site remediation can quickly escalate into considerable sums given the complexities of subsurface conditions. The remediation plan and execution of the cleanup is generally the most costly and time-consuming portion of the project. As the cost of site cleanup has risen in the past few years, questions are being raised concerning what governing philosophies should be. Considerable progress has been made in regulation, engineering practice, environmental awareness, source control, and waste management procedures. When used with sound geologic investigation, this progress enables more reasoned decisions to be made regarding contaminant migration and threat to human health. The EPA and some states have prioritized sites for cleanup, so contaminated sites posing potentially extreme health hazards are addressed first.

Site Soil and Groundwater Cleanup Goals

Site soil and groundwater cleanup will depend on the cleanup standard for that site, which is usually derived to protect water quality standards. As contaminant extent becomes understood, site cleanup is really the combination of soil and groundwater cleanup goals. The soil must be remediated so that a continuing groundwater contaminant source does not remain. Soil cleanup goals have been developed by the states regarding petroleum fuels and some volatile organic compounds. A list of states' cleanup standards for soils and groundwater (as developed to that date) was published in Kostecki et al. (1995). While each state is moving toward establishing goals, these differ in the actionable contaminant levels, threshold cleanup levels, and compounds of interest in each state. The reader is referred to those lists and must review the individual state guidance for which the problem must be solved. Generally, the soil cleanup level is less strict that the groundwater standard and concentrations levels at- which actions must be taken vary.

Groundwater cleanup guidance is typically derived from the drinking water protection standards. Drinking water quality standards have been developed over the years by EPA and by California, as well as other states. Brief reviews of how these standards have been developed and which are enforceable are listed below:

EPA Maximum Contaminant Levels (MCLs). Also known as the Primary Drinking Water Regulations. Federally enforced, MCLs are set as close as possible to the health-based Recommended Maximum Contaminant Levels (RMCLs) after taking into account the best technology available and cost to achieve the standard for drinking water.

EPA Recommended Maximum Contaminant Levels (RMCLs). These are developed by the Office of Drinking Water as the first step in the promulgation of MCLs. RMCLs are strictly health-based, derived from toxicological data. RMCLs are health-related goals but are not enforceable as drinking water standards.

California examples — Health Advisories, previously called Suggested No Adverse Response Levels (SNARLS), which are derived from the National Academy of Sciences (NAS) and EPA information.

California Department of Toxic Substances Control (DTSC) Action Levels. Similar to the EPA and NAS health-based criteria, derived in the same way. The Action Levels are not enforceable in the same sense as MCLs but are levels at which DTSC requires water purveyors to take corrective action to reduce contamination in water supplies.

Site Cleanup Level Approach

It is very important to realize that complete cleanup of every molecule of contamination is impossible, and that some residual contamination will remain. An evaluation of the threat of this contamination will remain, so residual contamination can become a challenging issue of "how clean is clean" in negotiations. A related issue of aquifer protection is that of soil quality and its residual threat to groundwater. Often it might be somewhat easier to clean the soil/sediment than the groundwater. A cleanup guidance and standard will result from the agencies' review of the site investigation information. The preferred approach is to remove the source and peripherally contaminated soil and groundwater to the point where the contaminant is "not detected" by chemical analysis.

Occasionally, the regulating agency may use a risk-based calculation approach to "allow" a higher concentration of the contaminant in soil to what is proposed in water. This assumes that the soil will provide a buffer or attenuate the contaminant without leaching downward to the aquifer. Often, capping the site or affected area is required to minimize vertical leaching potential with remaining soil contaminants. Other issues that also come into play with groundwater protection are: type and thickness of sediment strata, total thickness of the vadose zone, type and concentration of contaminant, and present and future land use. Groundwater will be viewed from the position of not allowing drinking water to be degraded. The cleanup level may be set at the MCL if contaminants are present in a usable aquifer. If the contaminated groundwater is separated from the drinking water aquifer, levels may be set above the MCL, with long-term monitoring to ensure that concentrations continue to decrease.

Risk assessment is increasingly used to ascertain the attenuation and possible exposure of contaminants, and then used to help select the approach. Given the increasing cost of site cleanup for very low levels of residual contamination, this approach should be used. The contamination will never be "100% cleaned up"; however, it implies that regulating agencies may not "sign off," which states that the site is clean. Regulating agencies may reserve the right to reopen the case should site conditions change in the future.

OVERVIEW OF RESPONSE ACTIONS AND CLEANUP TECHNOLOGY

Technologies have been developed for site cleanup, and performance-tested and used at numerous sites throughout the nation. The usefulness of the technologies may depend on the contaminant type, site soil and sediment type, hydrogeology, size of the problem, and selected technology cost-effectiveness for human protection. When the contaminants have been located and defined, then they have to be removed and treated as effectively as possible to reduce or remove the health threat. The contaminant could be collected and recycled, destroyed or absorbed onto other material for subsequent treatment or recovery, and ultimately some portion may go to the appropriate permitted disposal as, and if, allowed by law. State and national regulations are aimed to lessen the amount of material ultimately removed for off-site disposal. To that end, EPA is testing various new and innovative technologies at sites and provides reviews of the effectiveness and usefulness to the public (see, for example, U.S. EPA, 1994b.)

There is no magic formula for site cleanup technology, and one particular technology and cleanup approach is not suitable for all sites. The technologies can be used separately or in tandem to attain site cleanup. For example, groundwater pump-and-treat is used with soil vapor extraction and air sparging to clean up the dewatered portion of the aquifer. Another approach is to hydraulically or pneumatically fracture tight silt or clay formations to allow vapor or water to migrate toward the extraction wells. Each site is different, and the combination and deployment of technologies and their modification to site-specific conditions is critical for site remediation.

Site geology and hydrogeology may greatly inhibit the remedial effectiveness with the currently available technology. EPA (1993) has prepared a "Technical Impracticability" guidance recognizing that some contaminants, such as DNA-PLS, may not be completely removed. Although proper site assessment and cleanup actions are selected, the operational remedial technology has its limits and it is very reasonable to entertain the option that some sites cannot be completely cleaned up. While this is not a substitute for not cleaning as much as possible, it does recognize the presence of certain intractable geologic and contaminant dispersal situations, providing that site-specific data supports presence of these problems. This has since been somewhat reinforced in the "nonattainment" concept (recently proposed for California groundwater basins) implying the acceptance of residual contaminants following the required remedial feasibility study or cleanup effort.

An implication here is that long-term monitoring is required for all sites, and monitoring to detect low concentrations of contaminants for some time after active cleanup. The periodic detection of contaminants must be discussed in the context of overall site conditions and geology, especially where the geology and hydrogeology constraints have limited cleanup effectiveness (such as low permeability materials and residual contaminant drainage). The argument for the duration of remedial action and what the monitoring data trends indicate should be kept in this context.

After site characterization is complete, the type of response needs to be decided. This step will be done in consultation with the agencies to select the alternative for the site needs. Some of the types of responses and actions that may be taken are listed below (see Norris et al., 1993; Testa and Winegardner, 1991; U.S. EPA, 1994a). It should also be noted that there are hidden costs in all projects; these can include regulatory oversight and review time charged to the client, operation and maintenance of technology and equipment, public meetings, adapting existing technology to sites with less than favorable conditions, nonacceptance of the lowest cost for anticipated effectiveness, and fear/politics generated by competing interest groups. Brief reviews of response action and cleanup technologies follow.

Site Monitoring

This action will be needed at each site to ascertain the effectiveness of the groundwater extraction and plume migration, and to confirm that additional contamination is not leaving the site. This includes chemical analysis for quarterly groundwater monitoring (and can include vapor monitoring as well as discrete subsurface and stockpile soil sampling), with reports send to the appropriate agencies. Once the site is considered "clean," then the site closure may be attempted where the closure decision implies a cessation of monitoring and remedial action with the understanding that should conditions change in the future, the agency may reopen the case. The duration of site monitoring is variable; however, it must show that the problem has been adequately dealt with and the remaining concentrations are declining, or not detected. The length of monitoring can vary widely, but for petroleum fuels it may be a minimum of 2 years (eight sampling events) to establish the declining or not-detected trend. Large industrial sites, or sites with very large plumes, may monitor years or decades after active cleanup is completed (such as RCRA 30-year postclosure). The financial commitment of the client to the duration of monitoring can be substantial.

Excavation

This is the physical removal of contaminated soil, either for off-site disposal or on-site treatment. Soil treatment could use bioremediation, incineration, aeration, or other technology performed with the required permits, operation needs, and landfill acceptance. Landfilling is discouraged by new regulations to prevent "moving the problem elsewhere" and although still an option, can be very costly. However, for some contaminants, such as petroleum fuels, soil can be treated to levels acceptable for landfilling or some type of reuse. At times, removal of contaminated material may be the only option if on-site treatment interferes with site operations. Postexcavation sampling of the pit is typically required before excavation backfilling to show removal of contaminant to the satisfaction of the regulators and site cleanup level.

Soil Vapor Extraction

This involves installing extraction wells to remove and exhaust soil vapor contaminated with the volatile phase of the contaminant. The contaminant volatility is used to both mobilize and move the contaminant to the treatment. Commonly used at fuel UST and volatile solvent sites, it is an efficient cleanup technology for soil, especially where the vadose zone is thick and porous (Figures 2 and 3). The extracted vapor is given on-site treatment to the permitted levels and discharged. This technology is relatively easy to design and use, with "off-the-shelf" components, is cost-effective for small sites, and is "proven" in reducing the bulk of contaminants and the BTEX components of fuels and volatile solvents (such as TCE, PCE). Site specific engineering is needed to ensure the maximum efficiency of the wells and vapor removal (radial well influence, well placement and volume extracted, treatment; see Johnson and others, 1990a, 1990b). Post-vapor-extraction discrete soil sampling for residual concentrations is usually required to ascertain removal effectiveness as part of the site closure process. As a very general rule, soil vapor may lower soil concentrations of TPH to below 10 ppm and BTEX to below 1 ppm.

Hydraulic Fracturing

This is not a cleanup technology per se, but is an aid in opening low-permeability formations (either vadose zone or aquifer) to allow fluid circulation. When used in a clay soil for soil vapor extraction, for example, a high-pressure airstream is blown into an injection well to force openings in soil and weathered

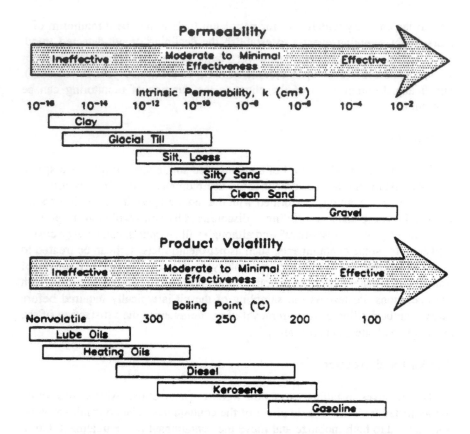

Figure 2   Initial screening parameters for soil permeability and product volatility to determine soil vapor extraction effectiveness. (From U.S. EPA, 1994a. With permission.)

rock. Sand may be blown in after the initial fracturing to hold the openings apart. Soil vapor extraction would follow the fracturing, and both the air extraction and volatile content of the airstream should increase. Similar procedures would be used to open formations that have low water yields, where again the overall purpose is to attempt to create more void space to increase remedial system effectiveness. Fractures may shift, reopen, or close with time depending on site geology. The overall effectiveness of hydraulic fracturing depends on the soil type or formation, location of buildings, contaminant type and extent, and degree of fracture closure over time.

Separate Phase Product Removal

This is an obvious first step if separate phase "free" product is observed in monitoring wells. The approach is to use any type of product recovery or bailing to collect and remove as much product as possible. The removal allows an immediate site response and lessens the quantity that could be ultimately dissolved in groundwater. As discussed elsewhere, the calculation of the quality of

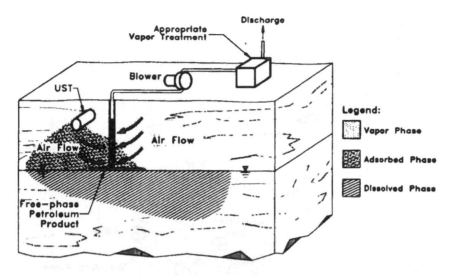

Figure 3 Conceptual operation of a soil vapor extraction system. (From U.S. EPA, 1994a. With permission.)

free product on the capillary fringe is speculative, and bailing tests of wells actually may yield the most useful remedial information. Soil vapor extraction has also proved effective in separate phase removal. Case histories have shown that enormous quantities of separate phase product may occur given the leak history and geology. Long-term removal of LNAPL or DNAPL may be factored into the groundwater containment and surface collection vessels needed for storage. When sufficient quantities of product are collected, it may be removed by recyclers.

Groundwater Containment with Subsurface Structures

This is not a treatment, but a technology to provide a physical capture of contaminated groundwater. Structures have been used for years for water barriers for deep construction and dam foundations (such as steel-driven piles, bentonite walls, or grout curtains). This approach could be used where the volume of water is very large, or with fixed sources such as landfills, or to protect nearby potable water wells. A subsurface wall is constructed by using low-permeability materials, such as bentonite clay or cement grout, to create a wall that is founded into a low-hydraulic-conductivity layer. Pumping wells may then pump against the walls for contaminant capture. Walls and barriers can be effective to gain time for cleanup but can be very expensive, and they require careful construction and monitoring for effectiveness. This type of containment is not usually financially feasible for small sites or clients with limited resources, since these costs are added to the water pumping and treatment. Walls may leak and move, so geotechnical engineering design and control as well as groundwater monitoring at the walls are often required.

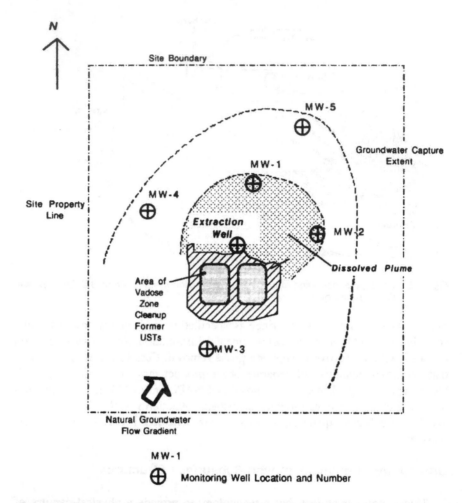

Figure 4 Conceptual of groundwater capture. The groundwater extraction well is located near the contaminant source and the pumping influence or extent is sufficiently large to encompass the dissolved plume. The capture area is elliptical and causes the depression, so the natural flow will be captured by pumping. The size of the capture extent may vary depending on gradient inclination and rate of pumping.

Groundwater Pump-and-Treat and Hydraulic Containment

This uses groundwater extraction for a surface treatment of the groundwater and provides the hydraulic containment of the dissolved plume. This can also include separate phase product capture and recovery (Figure 4). Contaminated groundwater and separate phase product are captured and extracted to the surface treatment. Monitoring wells are used to monitoring the pumping surface influence and collect periodic water quality data for system effectiveness. Once groundwater pump tests and characterization are complete, then modeling some different pumping well placement and yield scenarios might determine the useful number of pumping wells and their location (see Keely, 1982, 1984,

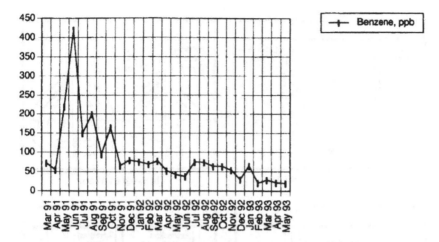

Figure 5   Data plot of benzene from a groundwater pump-and-treat system, March 1991 to May 1993. The largest removal occurs in the early stages of pumping and is followed by a declining trend. Fluctuations in the trend could be caused by contaminant drainage or seasonal precipitation and leaching.

1989). The groundwater pumping hydraulic contains the plume and flushes lower contaminant concentrations and uncontaminated water from the plume perimeter through the contaminated area. Although costly and long-term in use, these costs are low when compared to first installing a physical barrier followed by pumping.

Usually long-term trends of dissolved contaminant concentrations will decline to an asymptote, removing the bulk of contaminants, then slowly removing residuals by desorption. The question is, beyond what cost is further groundwater extraction not effective? This in turn depends on the contaminant type, duration underground, site geology and hydrogeology, and absorption and desorption trends of the contaminants. The long-term trend of contaminants continues to decline without increases (except perhaps short leaching episodes from seasonal rains) documented by monitoring (Figure 5). Hopefully this will occur near the regulatory site cleanup level.

Variations of pump-and-treat can include working in tandem with other listed technologies (i.e., sparging, soil vapor extraction, and physical containment) and with one or more pumping wells. At times both soil vapor and groundwater may be extracted from the same well and the entire extracted stream sent to surface treatment equipment. Newer technologies include horizontal wells where long "gallery"-like wells are installed to get a long area of influence, or in lower permeability sediments and settings. The use of horizontal wells may be of great use in areas of surface development, or large contaminated areas (see Wilson, 1994, 1995). A recent paper by Bartow and Davenport (1995) reviews the effectiveness of groundwater remediation in the Santa Clara Valley, California. Their conclusion is that pump-and-treat has been successful in significantly reducing volatile organic concentrations in groundwater, but has had limited success in removing dissolved concentrations to levels below MCLs.

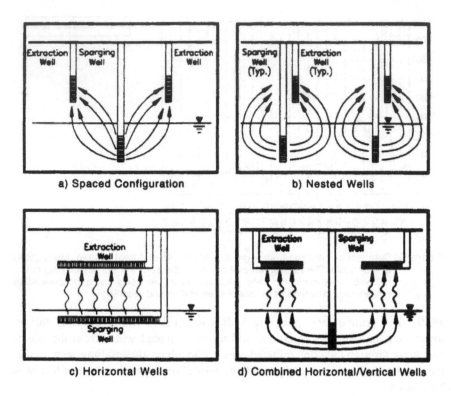

a) Spaced Configuration

b) Nested Wells

c) Horizontal Wells

d) Combined Horizontal/Vertical Wells

Figure 6 Conceptual arrays for sparging wells and extraction points; wells must be placed so their operation does not adversely affect plume movement. (From U.S. EPA, 1994a. With permission.

Air Sparging

This is a relatively new technology where air is injected into the subsurface and below the groundwater surface to assist in volatilizing the dissolved contaminants for recovery by soil vapor extraction. The air injection and movement assists volatilization in groundwater and movement to both soil vapor and water capture wells (Figure 6). The advantage is to help clean the aquifer and vadose zone at the same time by flushing air through the affected zones to speed cleanup. Once the site is delineated, several wells are installed and air sparging sources may be either placed in existing wells or separately. The use of air sparging and soil vapor extraction in tandem allows capture of the volatilized contaminant, and helps to extract immobilized contaminants in the aquifer to the surface for treatment. Problems include short-circuiting of air pathways, and the completeness of the sparging effect is doubtful in some cases. It requires porous and permeable geology in order to work, and site investigation and testing can include sparging, soil vapor extraction, and groundwater pumping tests for proper design. Reviews of air sparging, its use, and its limitations are presented by Dahmani (1994) and Ji et al., (1994).

Soil Flushing

This approach uses some technique (flooding, surfactants, steam injection, etc.) to remove soil contaminants. As the soil is flushed, the contaminants are mobilized and driven downward to the groundwater and are then captured and pumped to the surface. Care must be taken so the contaminants are not allowed to further migrate into vadose soil or groundwater and spread the problem beyond its original plume limits or the system ability to capture. This requires intensive geologic characterization and monitoring, and residual contaminant could remain. Site vadose zone geology should be very rigorously characterized, as should groundwater, and monitoring through the soil flushing process may be needed in addition to the routine monitoring.

Air Stripping

This is a surface treatment system where the volatile contaminant is vaporized by high-speed air forced through a tube while contaminated groundwater flows downward, stripping the volatile contaminant from the water. It is very efficient for certain volatile contaminants, has a long track record of use, and the technology is "off-the-shelf" and relatively easy to engineer, install, and maintain. However, air discharge permit requirements are an issue as air quality issues evolve. The option to remove the contaminant from groundwater to be released into air is becoming limited. When an air issue is a problem, then the airstream is treated a second time, sometimes called polishing, where the remaining contaminants are absorbed onto carbon or incinerated before being discharged at the regulatory approved levels.

Carbon Adsorption

This is a very widely used surface treatment for petroleum fuels and solvents. The ability of the large surface area of carbon to adsorb volatile contaminants is used to clean water or airstreams by forcing the contaminated medium through the carbon; given the required retention time for the type of contaminant, this removes the pollutant. The canisters may be placed in series so that if the contaminant breaks through the first canister, another is already in place to continue treatment, and the spent canister is removed and replaced. Carbon canisters come in many sizes, are easy to size for the contaminant load, can be shipped and moved easily, and spent carbon can be replenished or disposed as needed. This is very efficient, but can be costly depending on the contaminant concentrations and volume of carbon used. It is often used in tandem with air stripping, air sparging, and soil vapor extraction, and used extensively to "polish" (remove very low concentrations) in soil vapor or groundwater to meet permit discharge requirements.

Incineration

This is the thermal combustion treatment of contaminants destroying them. It may be used on a variety of contaminants, including domestic refuse. The efficiency of combustion is monitored through retention time for burning, depending on the type of contaminant and equipment used. The advantage is the complete destruction of the material, with inorganic ash that is then landfilled (a volume reduction in material landfilled, and an inorganic ash is somewhat easier to handle for disposal). Stack scrubbers and technology for monitoring the degree of combustion and aerosols monitor the operation for adherence to the required air permit. Once the contaminant has been incinerated, the client may receive a certificate of destruction, an added plus for reducing liability.

In Situ Stabilization

This technology uses stabilization techniques to prevent contaminants from leaching in soil. It may be used more in the vadose zone than in groundwater, and appears to be more useful for heavy metals than organic pollutants. Some type of fixing agent is placed in the contaminated area, which immobilizes the pollutants. This could include chemical fixing agents, cement, groundfreezing, and waxes. Other recent technology being developed includes moving high-amperage current through soil to vitrify soil (essentially melting soil to a glassy mass), preventing future contaminant leaching and migration. This can be a very expensive process. If the contaminated material is excavated and treated above-ground, costs may decline; however, the treated material is sent to a landfill or secure storage area. The EPA has numerous field and demonstration projects operating with this technology, and the reader is referred to the Kerr Laboratory (Ada, OK) and EPA Technology Transfer (Cincinnati, OH; see, for example, EPA, 1994b).

Bioremediation

This "technology" involves indigenous microbiota that use the contaminant as an energy source, and yielding respiration products of carbon dioxide and water. Naturally occurring microorganisms are supplied with oxygen and mineral nutrients (such as nitrogen and phosphorous) to aid in microbial growth in the contaminated area. This has been demonstrated to work for petroleum fuel contamination of soil and water, and is used extensively for surface soil cleanup. Recirculating groundwater pumping systems can be used in favorable geology for in situ aquifer cleanup. Surface "passive bioremedial" work tilling soil and adding nutrients and water to the increased soil area for hydrocarbons degradation can be very cost-effective. Another variant, bioventing, is used where air is blown through soil piles or the subsurface, enhancing the available oxygen to aid microbial growth (Figure 7).

Bioremediation has been used in the water treatment field for years, and use on petroleum contaminants has been done since the late 1960s. It is effective for

FAVORABLE FACTORS

CHEMICAL CHARACTERISTICS
small number of organic contaminants
nontoxic concentrations
diverse microbial populations
suitable electron acceptor condition
pH 6 to 8

HYDROGEOLOGICAL CHARACTERISTICS
granular porous media
high permeability (K > 10⁻⁴ cm/sec)
uniform mineralogy
homogeneous media
saturated media

UNFAVORABLE FACTORS

CHEMICAL CHARACTERISTICS
numerous contaminants
complex mixture of inorganic and organic compounds
toxic concentrations
sparse microbial activity
absence of appropriate electron acceptors
pH extremes

HYDROGEOLOGICAL CHARACTERISTICS
fractured rock
low permeability (K < 10⁻⁴ cm/sec)
complex mineralogy
heterogeneous media
unsaturated-saturated conditions

Figure 7 Favorable and unfavorable chemical and hydrogeological factors for in situ bioremediation. (From U.S. EPA, 1994a. With permission.)

light, nonchlorinated, and lightly chlorinated hydrocarbons. This treatment technique is widely used today since it treats the problem on site, degrades the contaminants to acceptable levels in days to weeks (depending on weather and initial concentrations), and allows the soil or water to be sent to less costly disposal, such as recycling for fill, or to an asphalt plant for use in asphalt aggregate or roadbase. The reader is referred to Norris et al. (1993) for a review of the chemical and geologic considerations for bioremedial actions.

Natural Attenuation

Natural attenuation is the process by which the contaminant is naturally degraded in the subsurface, and site monitoring tracks the declining concentrations in site wells. As monitoring proceeds, the declining concentrations are observed to ensure that human health goals are met and that residuals continue to decline. This approach is very useful for petroleum fuels and the BTEX compounds that typically present the most immediate threat to human health when dissolved in groundwater. Hydrocarbons are known to naturally degrade with time in aquifers (Norris et al., 1993).

To briefly summarize Norris et al., the hydrocarbon compounds will degrade through aerobic and anaerobic reactions where the microorganisms utilize the hydrocarbons as an energy source. Almost all hydrocarbons are degradable in aerobic conditions. Factors that limit degradation are oxygen, available nutrients, temperature, and pH. When a contaminant plume enters and moves downgradient with the groundwater, aerobic conditions will commence until the available oxygen is used. The plume migrates and mixes with more groundwater with higher oxygen contents and the biodegradation continues on the outer portions of the plume somewhat faster than on the plume interior (Figure 8). The process continues with time depending on the geologic environment type and can be enhanced if adequate nutrients and oxygen can be delivered to the affected portion of the formation. This implies more potential success in sandy strata than clayey strata, although the degree of success and degradation are site-specific to source, geology, and duration of problem.

As a general rule, the BTEX components of the plume migrate somewhat faster than the bulk of total petroleum hydrocarbons. Also, benzene and ethylbenzene tend to degrade more quickly than toluene and xylene (see Barker et al., 1987). As BTEX degrades over time, concentrations will asymptote in the range of parts per billion to not detected. The long tail-off of concentrations is partially due to the degradation reactions and residual drainage of contaminants into and migration through the aquifer.

Risk Assessment

This study involves a review of data to ascertain whether the risk is acceptable to leaving contamination on site or using a specific approach for site cleanup. This includes the threat to soil and rock, surface and subsurface water, habitat, industrial, agricultural, recreational, watershed, and other uses of soil, water, and land. It can include the dose study (ingestion of contaminants to calculate the lower doses acceptable in air, water, soil, etc.). The assessment is based on the exposure and release pathways of the pollutants and the amount of exposure people may have. It can be a part of the remedial action or site-closure plan, or be a stand-alone document. Risk assessment-based site cleanups and closures are being used in Oregon, Texas, California, and other states.

The assumptions made in the risk models are based on available health data and site-specific data, and use of realistic numbers is needed. As with other modeling techniques, assumptions and data manipulation can be used to support or refute positions. Risk assessment will probably be used more in the future to rationalize and support ending active remedial efforts; negotiation with regulators and sometimes the public is always implied in the risk assessment process. Recently developed cleanup standards and risk assessment procedures in Oregon and Texas use approaches flexible to attaining the cleanup without compromising protection, based on the site-specific conditions, contaminant type and concen-

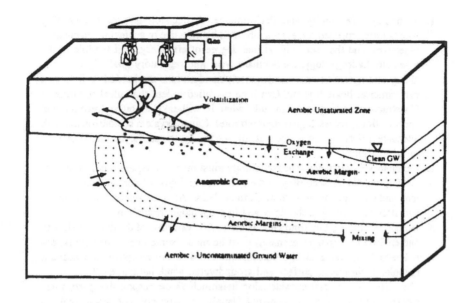

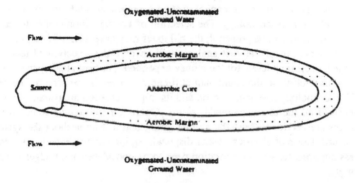

Figure 8 Conceptual of a groundwater petroleum hydrocarbon plume undergoing natural bioremediation processes. (From Norris, 1993. With permission.)

trations, and proposed or projected land use. Assessments may also be used in conjunction with deed restriction to property if certain levels of contaminants are left on site (and natural attenuation approach). Site monitoring and natural attenuation as discussed above are used to support the residual concentrations presence and threat, if any. When risk assessment is used with the proper site assessment approach and appropriate remedial action, it should assure human protection — although there is always some degree of risk, regardless of the protection afforded.

SELECTING APPLICABLE REMEDIAL ALTERNATIVES

Four primary factors are usually reviewed when selecting and considering site corrective action:

1. Technology Feasibility and Performance. A decision is needed regarding whether the proposed remedial action will be effective in achieving the cleanup objectives, and the useful life at that degree of efficiency. Is it feasible to the given site hydrogeology, contaminant type, and regulatory goal? System reliability and operation and maintenance considerations are needed such as past performance, flexibility, etc. Can it be modified to less than ideal conditions? Construction requirements include what it will take to build (permits, bid specifications, access, logistics, timeframe). Can the system be constructed with adequate safety, or even used at all — as, for example, incinerating in a residential area?

2. Regulatory Compliance. The technology must meet the requirements of the site cleanup objectives according to the guidance and type of contamination problem, and the negotiation of "how clean is clean." Are the cleanup levels realistic and attainable by the technology and hydrogeologic limitations?

3. Human Health and Environmental Factors. An evaluation of the threats to health and environment (risk assessment) must be made (sometimes called the potential receptors). Have all the considerations taken the receptors into account (people, environment, surface and groundwater, wind, habitat, etc.?

4. Cost. This is a crucial consideration inasmuch as the responsible party must pay and not all costs may be apparent. Usually some type of cost-benefit analysis will be performed. Initial capital costs are associated with review of the data and selection of technology. The selection of a remediation contractor and bid specifications will be prepared; the bid itself may have a significant preparation cost. A portion of the cost analysis will involve the selection of technology based on cleanup levels for the money expended. The best cleanup may become cost-prohibitive to the client, and at times may even be resisted by the public. The technology selected, its cost, and its implementation are usually a compromise of all the aforementioned factors. Operation and maintenance costs and plans are needed to keep the system going and to troubleshoot the system as needed. Costs of electrical, water disposal, equipment breakdown, and realistic assumptions for smooth operation need to be considered and budgeted accordingly.

Duration of Cleanup and Post-Cleanup Monitoring

A question of how long the treatment–cleanup system must stay in operation is always present at the beginning of work. The length of time is always an estimate, based on the assumptions that the system and approach will work as envisioned with minimal problems and most importantly, the site cleanup concentrations levels. The cleanup phase can easily be much longer than the investigation phase. For example, excavation and removal projects may run into weeks and months if a large quantity is involved, followed by postcleanup sampling and reporting. Soil vapor projects may be months to years long. Removing groundwater contaminants may range in years or decades when active treatment and postcleanup monitoring are factored in. At times the client may wish to accelerate the work if money or other compelling reasons exist, and may diminish cleanup time. This "brute force" approach is effective, but the option to spend large quantities of money quickly is rare, and ultimately the time factor is most often related to remedial cleanup efficiency in reaching the mandated cleanup concentrations.

SCREENING THE APPLICABLE REMEDIAL ALTERNATIVES

The alternatives must be screened on the basis of environmental and public health criteria, technical merit, and cost. The screening process should eliminate alternatives which are not appropriate and do not adequately provide the cleanup required. A rationale for the selection of deletion of each alternative is usually given and leads to the final selection.

1. Environmental and Public Health Screening. This identifies the adverse impacts upon the receptors, and the technology must address the relationship of the site and contaminants to potential exposure pathways, define potential receptors, estimate exposure concentrations, and give the anticipated timetable for cleanup.
2. Technical Justification. Alternatives are based on technical feasibility, site geology, engineering and system design for the site, aquifer characteristics, groundwater flow rates, soil contamination, contaminant types and concentrations, and so on. Alternatives are then compared and eliminated or accepted.
3. Cost Factors. Cost screening is performed to eliminate costly designs and to estimate the cost of the cleanup. Depending on the anticipated efficiency of the alternative, comparisons to other costs are made. A typical approach is to look at costs and choose the one that will perform reliably and do the job consistently. Cost considerations are: permits (right-of-way or easement access, National Pollution Discharge Elimination Standard permits, Publically Owned Treatment Work permits, local fees) and regulatory charges for permits and document review charged to the client, engineering and construction costs, start-up, operation and maintenance budgets, contingency, monitoring, chemical analysis budget, reporting and site-closure costs.

IMPLEMENTATION OF CORRECTIVE ACTION

Once the remedial alternative has been selected and approved by the parties, it can be implemented. The selection criteria may have been approved in the Remedial Action Plan report, and even commented upon by the public. Coordination between the site manager and consultant is worked out and all parties are notified as to when the system will start, permits are in place, etc. The monitoring plan should have been implemented with periodic reporting.

The system is installed and started up to determine that it works properly, and that the initial performance is as anticipated. When new systems are installed, unforeseen problems can and do crop up, and the site manager and contractor must address them to attain the cleanup standard. System modifications should be made when the system operation indicate and can enhance performance. Finally, the verification monitoring and sampling of the remedial action must be done. This includes the sampling of vapor, soil, and groundwater to ascertain the effectiveness of the system and to observe the decrease in contaminant. Over time, the track record of the system performance is recorded in the periodic reporting and monitoring. Once the cleanup goals are obtained, then the site closure may proceed. This may include a separate report or letter to the agencies, and may require a final sampling and analysis.

FINALIZING THE CLEANUP PLAN

Typically, the following steps are negotiated and resolved prior to starting the final cleanup. Any number of additional steps could be required by site conditions or regulatory order. Public release and meetings for discussion could be required at various stages, and a public awareness program may be a part of each step. These steps represent typical parts of the remedial action plan, as well as contractual or budget considerations listed below.

- Final demarcation of the area to be remediated, the zero contamination (or none detected) line. Failure to locate the edge of cleanup, and agree to that boundary with the regulators, may result in additional investigation and cleanup. Additional hydrogeology and engineering costs may arise following the initial site investigation, to collect more information if the assessment is incomplete.
- Acceptance of the cleanup plan by the regulating agencies (with public review if required). This will involve complete negotiation of contamination extent, cleanup methods, cleanup costs, and selected cleanup technology and related engineering design. Resolving these issues may take years and may require substantial changes in the technical approach (even possibly new investigations).
- Securing the needed permits from the appropriate agencies. This may include excavation, equipment and well construction, waste discharge methodologies, and permits.
- Preparation and review of the site safety plan and personnel medical monitoring as required.
- Determination of the costs of cleanup. This varies depending on the remediation method and can include budgets for cleanup, maintenance, monitoring, and site closure. Cleanup costs are typically high, easily reaching the tens or hundreds of thousands of dollars (not including future monitoring or other efforts). Usually a cost analysis is included in the cleanup plan when the most cost-effective remedial option is determined. Cost may change due to inflation, needs of additional equipment needs, additional review by regulators, effectiveness of the initial cleanup, or changes in subsurface conditions not anticipated in the site investigation.
- Selection of the cleanup contractor, usually by competitive bid. The bid proposal would include the cleanup plan and documentation so the contractor will perform the work in a cost-effective manner and negotiate additional funds for unforeseen contingencies or problems.
- Finalization of any physical plant or containment facilities, including design engineering and drawings, work timetables, safety plan, and other site procedural documents.
- Set up the decontamination area for field work; or construct and install the needed treatment, monitoring, and safety equipment (including the calibrating and adjusting equipment, long-term maintenance, etc.).
- Verification monitoring or sampling to evaluate the system's effectiveness for site cleanup. This would include interval reporting (usually quarterly) following system startup and preparing reports for regulatory oversight.

A remedial action plan or site-closure plan is usually required at to detail the aforementioned points to the agency. An example of a remedial action/closure plan format and components are;

- Background history
  - cause and location of contaminant release
  - how release was detected
  - estimate of duration and volume of release
  - type of leak-detection system installed at site
- Site characterization
  - subsurface exploration and soil sampling methods
  - groundwater monitoring well design
  - groundwater sampling methods, water-level measurements
  - sampling protocol (analytical chemical lab, chain-of-custody documentation, sampling preservation, etc.)
  - regulatory requirements for analyses performed
  - methods used to detect and measure separate phase product
- Extent of soil and groundwater pollution
  - vertical and lateral extent of subsurface contamination
  - number and location of exploratory borings
  - number and location of monitoring wells
  - definition of separation phase product
  - definition of dissolved contaminants
- Hydrogeology
  - subsurface lithologies, primary and secondary permeability
  - aquifer characteristics
  - aquifer and aquitard relationships
  - groundwater flow direction and gradient
  - seasonal and diurnal groundwater elevation changes
  - geologic cross sections
  - preferred contaminant pathways
  - permeability characteristics of the vadose zone
- Beneficial uses of groundwater
  - existing and future groundwater uses
  - Groundwater Basin Plan requirements
  - potential receptor/risk assessment
  - ability to attain the cleanup levels; presence of preexisting water problems
- Remedial action
  - interim remedial actions used
  - development of remedial alternatives
  - screening remedial alternatives/technology/engineering
  - rationale for selected remedial action
  - soil remediation method (excavation, vapor venting, etc.)
  - groundwater remediation method
  - potential/existing impact of remedial action(s)-risk assessment
- Effectiveness of remediation
  - are cleanup levels consistent with federal/state guidelines?
  - verification monitoring program
  - potential impacts from residual contamination

- Site closure
  - closure plan reviewed by lead agency
  - verification monitoring data reviewed
  - "sign-off" (no further work required; cease monitoring)

EXAMPLE OF ATTAINING CLOSURE OF A "TOXICS" SITE

Problem

A large electronics firm had used a commercial building as a research lab for developing printed circuitboard manufacturing. The site was located in a business park that had other similar firms and manufacturing in the region. The firm used the site only for research and development, not for any manufacturing operations. The site used some solvents in the laboratory, but had kept inventory control and containers in the building. An exterior sump was used on a limited basis associated with the printed circuitboard operation, but did not receive solvent waste. When the firm decided to move, they performed sampling under the building slab to ascertain whether metals and organic contaminants were present. One soil sample detected TCA, toluene, and benzene. One groundwater monitoring well had been installed just downgradient of the sump and sampling revealed that TCA, 1,1-DCA, 1,1-DCE, PCE, TCE, and Freon 113 (solvents and compounds commonly used in electronics manufacturing) at or above state drinking water standards. Only TCA and Freon 113 had ever been used at our site and no other firm operations used chemicals at this location. Since the groundwater monitoring wells showed organic chemicals considered toxic, the regulatory agency wanted proof that the firm had not introduced the chemicals and they could not close the site until this had been done.

Problem Approach

The approach to this problem was twofold. First, only one groundwater monitoring well had been installed; in order to ascertain flow direction and confirm water quality, an additional two wells needed to be installed. Second, the site history and records and interviews with site personnel showed that the additional chemicals (TCE, 1,1-DCE, TCA 1,1-DCA, and PCE) were never used in site work. These chemicals also showed that these were solvents with possible degradation products (DCE and DCA) present, and indicated that another source of groundwater plume could be nearby. The documentation proof is on our site to show the problem is not of their making.

The additional wells were installed under an approved workplan and permit, and when sampled showed presence of 1,1-DCA, 111-TCA, PCE, 1,1-DCE, TCE, chloroform, Freon 113, and Freon 123A. Clearly, additional solvent-type organic compounds were present in the groundwater that had not been used at our site. Plots of groundwater gradient showed that groundwater flow was directly under the building, past the sump toward the far property line. A review of regulatory files showed that several known groundwater contaminant plumes occurred upgradient of the site, some of which were large and contained 1,1-DCA, 1,1,1-TCA, PCE, 1,1-DCE, TCE, and chloroform. They were also possible, but not confirmed, sources of Freon 123A. Additional checks of the records showed that the site was more than one mile from an open waterway, and no water-producing wells were within one mile downgradient of the site (Figure 9).

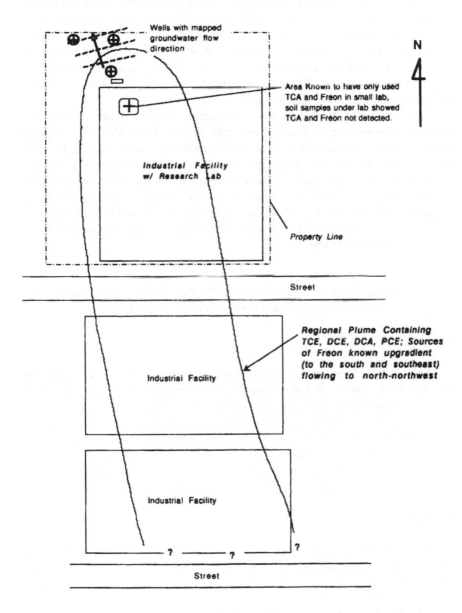

Figure 9 Regional plume proximity to site showing groundwater flow and lab location.

A report was prepared to document the additional upgradient sources of the organic chemicals, and with the site historical record, was sufficient proof of another source. The additional wells allowed the water flow direction to be plotted and showed the pathway toward the site. The concentrations were of values similar to those expected of a leading-edge plume, and contained degradation products expected to be observed from compounds that arose from off-site sources. It allowed for a year of groundwater monitoring with a minimum number of wells that the state agency felt was needed to show that the on-site sump was not a problem. Perhaps most importantly, the client had kept good records of the site use and chemical use history, and had performed sampling at their expense to document site conditions after their use. This was vital supporting documentation that their site was not a problem, and kept them from being drawn into a responsible-party battle to clean up a plume they did not create.

Upon review of the report, the state agency issued a letter allowing the cessation of monitoring, effectively the "site closure" letter, allowing the client to move. The report had all the minimum basic elements; site history and assessment, soil and groundwater quality, groundwater monitoring, supporting rationale of separate chemical problems and tracked to known sources of the contamination, and supporting documents for the state to review written in clear, unambiguous language. This also assisted the client and the real estate agent, since this contaminant problem was finished and the site could be sold or leased with the supporting documentation.

EXAMPLE OF SOIL VAPOR CLEANUP FOR CLIENT WITH LIMITED FUNDS AND MARGINAL SITE CONDITIONS

Problem

An industrial equipment yard had two fuel tanks used to refuel equipment that had been used for at least 25 years. Upon testing the pipelines and tank, leaks were discovered; the tank and contaminated soil were removed, and the tanks replaced. The client had only limited funds following the initial work due to a period of economic downturn. Although an investigation had to be performed and the regulating agency required action for site cleanup, the client had limited financial resources. The agency required that at least the minimum work be done to address the problem according to the guidance for underground tank leaks.

Approach

The approach was to investigate the site using shallow soil borings (to about 30–35 feet) to determine the extent of the vadose soil contamination. Soil sampling results revealed total petroleum hydrocarbon as gasoline (TPHG) with benzene, toluene, ethylbenzene, and xylene (BTEX) concentrations of 10 to 15,000 ppm (average) in a soil plume in the vicinity of the tanks (Figure 10). One groundwater monitoring well was installed since groundwater was 80 feet

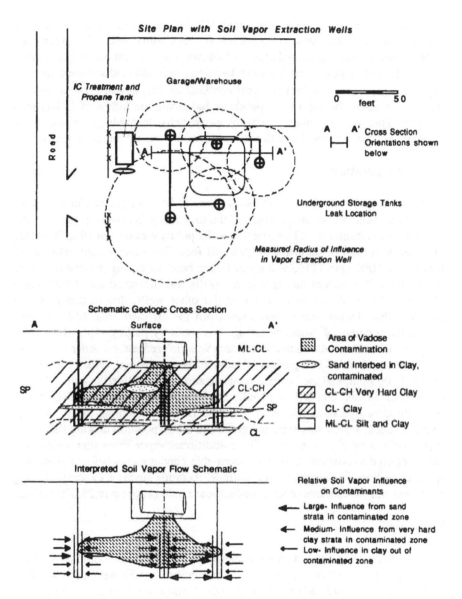

Figure 10 Soil vapor extraction from leaking tank site.

deep, and sampling showed that soil contamination ended about 35 feet below the surface and showed groundwater had not been affected. The site was underlain by a medium to highly plastic consolidated clay with thin sandy interbeds, and the apparently low rate of leakage over the years had been wicked into the clay. An initial estimate of the fuel leaked ranged from 15,000 to 25,000 gallons (the total amount was unknown since on-site inventory control was lax).

The cleanup approach selected was soil vapor extraction, which could remove the TPHG and BTEX without digging up the yard (thereby causing more financial harm by interfering with site operations). While it was conceded that the clay soil type was not favorable to vapor extraction, this would at least allow a remedial effort to run, protecting groundwater and allowing site operations to continue so the client could make income. Five well locations were selected and the extraction began using a surface portable internal combustion (IC) engine, with a metered fuel mix to allow the engine to operate by balancing the inflow of TPHG with propane. This was a proven, relatively inexpensive technology for small-site cleanup, and the permitted outfall was approved by the air monitoring board.

System Operation

The site operation of the extraction proved to be more successful than originally anticipated due to the presence of the thin sandy interbeds in the clay and contaminated plume area. The extraction wells pulled vacuums of 19 to 30 inches of water with radii of influence of 20 to 40 feet. The vapor stream into the IC was almost 1000 ppm TPHG and about 1 ppm benzene during the first 6 months of operation. The area of influence of the wells was measured and the influence of one well was almost twice as far as the other wells, due in part to sandy interbeds that thicken and extend away from the affected area. Although this caused the problem of "short-circuiting" the wells by pulling vapor from areas away from the plume, the contaminated area was covered in total. After 15 months of operation, about 13,000 gallons of fuel had been removed, with TPHG down to about 100 ppm, and the benzene concentration had declined to below 10 ppb (TEX constituents similarly declined). Exploratory soil borings and sampling collected after operation showed TPHG and BTEX remaining, locally at high concentrations, which was expected given the soil type and knowing that some vapor wells were short-circuited. Three additional vapor extraction wells were then proposed to continue to remove accessible contaminants before halting soil vapor extraction. The system would continue operation since vapor was still being recovered and the estimate of total product loss could range up to 25,000 gallons.

Epilogue

Admittedly, this was not an ideal approach, and the site conditions were not favorable to highly efficient removal rates of contaminant. However, this was also all that the client could afford. Even though cleanup was mandated by the state enforcement agency, the ability of the client to perform without becoming bankrupted must be kept in perspective. The approach and operation showed the extent of the soil plume, showed that groundwater was not impacted, and had removed almost half of the initially calculated loss. More important, the more mobile BTEX components had been significantly reduced, and the less mobile residual hydrocarbons lessened a future threat to groundwater. It is valuable to keep in mind that many factors bear on the ability to perform cleanups, not the least of which is the client's bank account. Many times, site conditions are less than ideal and ultimate removal of contaminants are less than everyone desires. The consultant often deals with a difficult interplay of these factors, and compromise is sometimes necessary to make the best of a bad situation.

EXAMPLE OF CLOSED-LOOP SUBSURFACE BIOREMEDIATION GROUNDWATER CLEANUP

A spill of about 1000 gallons of gasoline occurred at a service station site in the central coast region of California (see Figure 11). The portion of the spill retained in the vadose zone was excavated when the station was rebuilt and subsurface storage tanks replaced. An investigation was conducted to finish groundwater contamination definition. Initially, air stripping was proposed for groundwater remediation, but this was vetoed by the local air quality board. Consequently, a closed-loop subsurface bioremediation approach was suggested (Figure 12). The key to successful bioremediation is a porous and permeable aquifer that allows water to be continuously withdrawn and reinjected in order to deliver water and nutrients to the contaminated area. Site geology proved to be dune sand and marine terrace deposits, which are porous and permeable, and are suitable for bioremediation (see Figure 13).

Problem Approach

Since the California Regional Water Quality Control Board (RWQCB) needed to approve subsurface injection of contaminated water, additional aquifer delineation was required to show that it was separate from the regional sole-source aquifer, providing the only source of drinking water (Safe Drinking Water Act, 1974). Dissolved contaminant levels ranged from 10 to 30 ppm of TPHG and 100 to 200 ppb benzene (the most sensitive contaminant). Benzene concentration cleanup levels were negotiated to 7 ppb, (ten times the then required Action Level of 0.7 ppb) because the contaminated water was located in the perched aquifer system.

Additional investigations showed that the groundwater occurrence beneath the site was only semiperched and discontinuously in connection with the underlying sole-source aquifer (see Figure 13). Soil samples were also collected to model growth of the indigenous microfauna that would be used for the cleanup, and to estimate the quantity of gasoline in the aquifer porosity. An aquifer performance test was conducted to ascertain the optimum yield of the extraction well, observe the extent of the cone of influence of the well, and use the data to estimate nutrient-charged water injection rates. A step and constant-discharge test was performed, which determined that a well yield of 13 gallons per minute would create a cone sufficient to capture the contaminant plume (see Figure 14). Injection rates and resulting mounds were modeled for the optimum reinjection rates to stay within the cone and not disrupt the capture symmetry. The extracted groundwater would then be charged with the required nutrients and oxygen (as peroxide), reinjected around the periphery of the cone, and returned to the extrac-

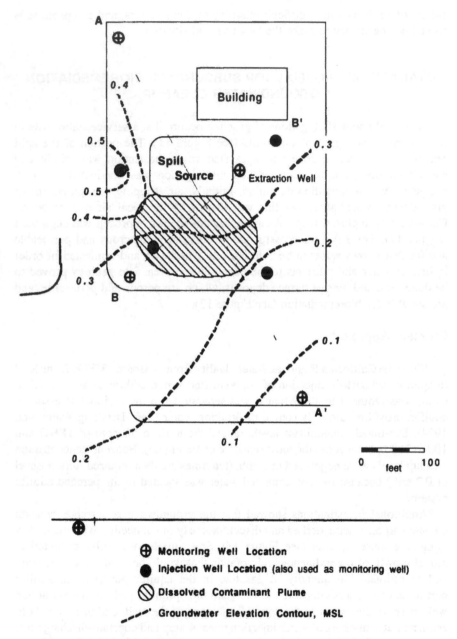

Figure 11 Site plan with contaminated plume, well locations, and pre-pump test groundwater contours.

tion well. The system was plumbed together and water recirculated to begin the
remediation. The original cleanup time was estimated at 18 months based on
computer simulations of extraction and reinjection to hold the contaminant plume
in place, and rates of microbial growth into the contaminated area.

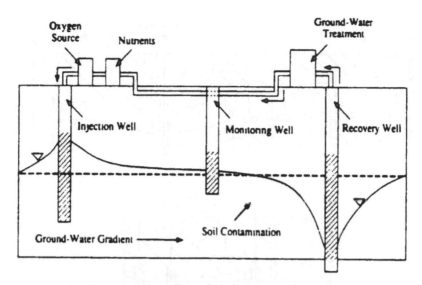

Figure 12   Conceptual of a subsurface closed-loop in situ bioremediation in the aquifer. (From U.S. EPA, 1994a. With permission.)

System Startup, Modification, and Effectiveness

System modifications were required during startup to balance flow and limit siltation of the injection wells. Both filters and well redevelopment were needed to remove silt to keep the flow at the design rates. After 6 months of operations, the contamination levels began to decline, and monthly declines were observed thereafter. Site monitoring was rigorous since the plume location, creation of phenols (from peroxide as an oxygen carrier) and free product release from microbial surfactants were requirements of the RWQCB. Frequency of monitoring decreased from weekly for the first month, to monthly and finally quarterly after the first three months of operation. Startup problems included the release of gasoline from the microbial activity, which was captured by the extraction well, and infilling of the injection wells due to silt content of the aquifer. The siltation of wells became a significant problem until wellhead filters could be installed. However, this additional maintenance caused budget over-runs for site operations.

The system was shut off during a regional aquifer water-level decline below the base of the monitoring and extraction wells. Once groundwater had recovered in the fall, the system was restarted and adjusted. A typical pattern of injection and extraction is presented in Figure 15. The site owner decided to accelerate the pace of nutrient delivery to speed up the cleanup. After one year of operation, the concentration of all contaminants had declined to none detected and the system was turned off. Monthly monitoring was continued for one year to assure that the cleanup was complete. The total cost of the site investigation, testing, system operation, maintenance, and monitoring to the point where the system could be shut down was roughly $270,000 in 1989 dollars.

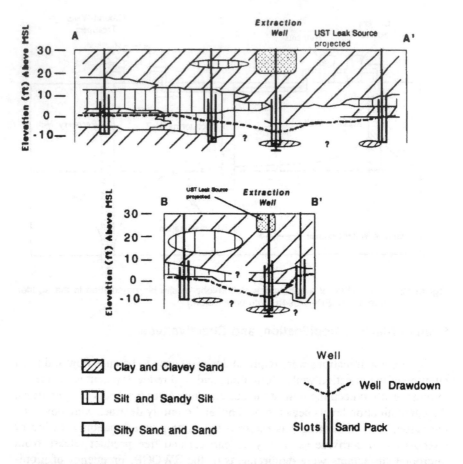

Figure 13 Geologic cross sections of stratigraphy beneath bioremediation site.

Bioremediation was effective for cleanup of petroleum hydrocarbons contained in the aquifer at this site. The key to success is the ability to recirculate groundwater in a pump-and-injection system. The nutrient and oxygen delivery rates must be matched to the indigenous microfauna ability to metabolize the contaminants. Unless water can be moved through the strata, the cleanup can be inefficient. The ability to bioremediate at this site allowed a cleanup of dissolved contaminants, and contaminants in the aquifer matrix, which protected the underlying sole-source aquifer.

VADOSE SOIL CLEANUP REMEDIATION PLAN DEVELOPMENT FOR A 2, 4-D SPILL (AFTER BLUNT, 1988)

Problem

In 1978, a large surface spill of 2,4-dichlorophenoxy acetic acid (2,4-D) occurred at a bulk transfer and product formulation facility in central California.

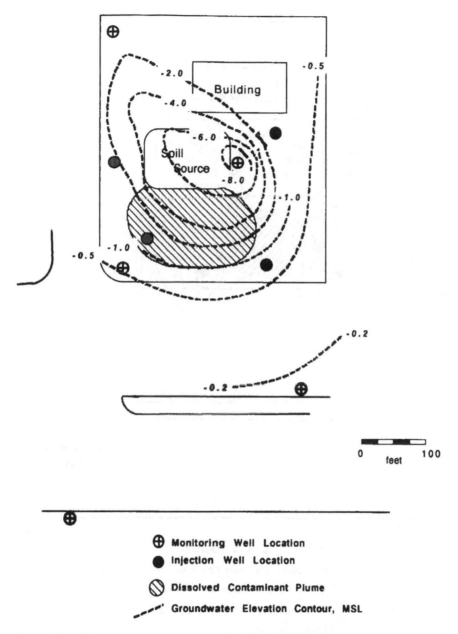

Figure 14   Observed drawdown after 24 hours of pumping at 13 gallons per minute.

Several phases of subsurface exploration (36 exploratory soil borings) defined the liquid 2,4-D spill penetrated the shallow vadose zone sandy and gravelly sediments to depths of 10 to 20 feet. Site groundwater monitoring wells showed that groundwater was not affected. The exploratory soil boring and sample analyses program ultimately defined the extent and depth of penetration of the con-

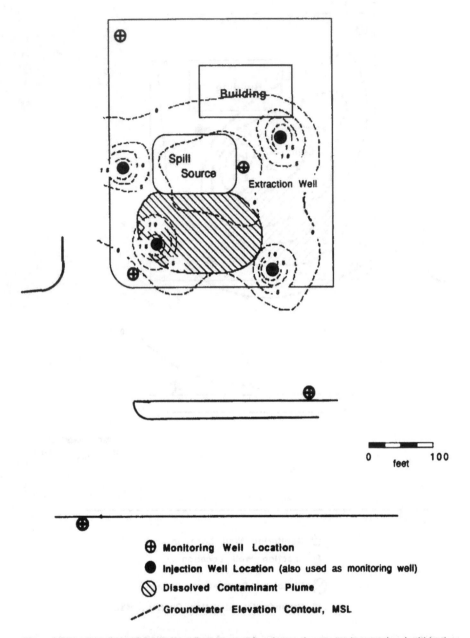

Figure 15 Injection and withdrawal contours after 8 months; mounds contained within the capture of the extraction well.

taminant by concentration (see Figure 16). In order to clean up the spill, a remedial action plan was needed for a cost-effective method of 2,4-D removal that would minimize exposing the surrounding residential areas.

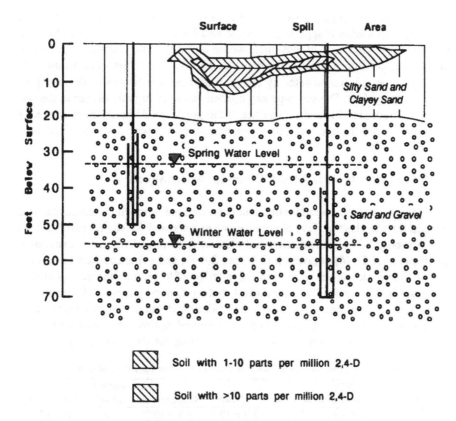

Figure 16   Soil contaminated with 2,4-D. Site hydrogeology definition showed that the contaminants were present only in the shallow soil and above the seasonal high groundwater level. Sufficient silt and clay were present in soil to inhibit vertical leaching.

Exposure Risk and Contaminant Pathway Identification

When the pesticide was spilled, it became a waste that state agencies regulated at 100 ppm for waste identification, 2,4-D is known to have toxic effects on humans at acute, subchronic, and chronic levels. Stomach ulcers and animal birth defects may be caused by 2,4-D and it is a potential (but unproven) human carcinogen. Consequently, the health risks of this compound need to determine the likelihood of injury from human exposure pathways. A judgment of the margin of safety had to be determined from the possible exposure routes. Since human exposure routes include dermal, inhalation, and ingestion, an evaluation of possible human exposure concentration and cancer risk were required to assess tolerable carcinogenic risk.

Pathways of exposure from 2,4-D were reviewed given the chemical characteristics of the compound. The compound has a low vapor pressure and movement by vapor diffusion would be negligible (Jury et al., 1983). Although 2,4-D is unlikely to be transported as vapor, it could be transported as dust. It could migrate by direct application if applied as a liquid and by leaching into the soil. Once in the soil, the mechanisms of movement would be by mass flow, liquefied diffusion, and gaseous diffusion (Hern and Melancon, 1986). Once in the vadose zone, 2,4-D may reach the groundwater by leaching and be drawn into wells supplying water to the public. The soil type and concentration of organic matter in the soil influence both mobility and persistence of the pesticide. Although pesticides may leach downward in soils, it is a slow process and moves pesticides short distances (Norris, 1966).

Developing Remedial Action Cleanup Alternatives

Once the vadose contamination had been defined, remedial alternatives were needed for site cleanup. Five alternatives were reviewed for site remediation: (1) monitor only; (2) bioremediation and enhanced bioremediation; (3) excavation to 10 ppm 2,4-D; (4) excavation to 1 ppm 2,4-D; and (5) excavation to <1 ppm 2,4-D. These alternatives were explored in terms of lessening the threat of human exposure and removing the contaminant. Safety factors, duration of the remediation, and cost for each option were also calculated and summarized (see Table 1). The cost for exposure and safety are weighed against the cost for cleanup; this information was discussed at a public meeting for the interested parties.

Table 1　Summary of Parameters Used to Evaluate Cleanup Alternatives

Alternative	2,4-D Concentration	Material[a]	Dose[b]	Cancer risk[c]	Safety factor[d]	Cleanup cost[e]
1	1600 ppm	Soil	2.3×10^{-3}	4.5×10^{-10}	1.3	$72,400
	100 ppb	Water	3.1×10^{-3}	6.1×10^{-10}	0.9	
2	100 ppm	Soil	1.4×10^{-4}	2.8×10^{-11}	21	$1,158,000 to
	110 ppb	Water	3.1×10^{-3}	6.1×10^{-10}	1.1	$720,000
3	10 ppm	Soil	1.4×10^{-5}	2.8×10^{-12}	210	$470,000
	10 ppb	Water	2.9×10^{-4}	5.7×10^{-11}	10	
4	1 ppm	Soil	1.4×10^{-6}	2.8×10^{-13}	2100	$600,000
	1 ppb	Water	2.9×10^{-5}	5.7×10^{-12}	100	
5	0(<1) ppm	Soil	$<1.4 \times 10^{-6}$	$<2.8 \times 10^{-13}$	>2100	$1,210,000
	0(<1) ppb	Water	$<2.9 \times 10^{-5}$	$<5.7 \times 10^{-12}$	>100	

[a] Ingestion exposure.
[b] Calculated for a 70-kg adult as mg/kg body weight/day by oral ingestion.
[c] Calculation based on a *pica* of a 1-day maximum exposure.
[d] The ratio of the MCL dose to the site calculated dose (0.1 mg/l × 2l/70 kg)/theoretical site dose 2,4-D)
[e] All costs assume a cleanup and monitoring program of 10 years at 1988 dollars.
From Blunt, 1988. With permission.

The "monitor only" alternative is the least costly and would not remove any contaminant. Blunt et al. (1988) calculated the time required for pesticide to naturally degrade to low levels is long compared to other alternatives, and provides a safety factor of 1.3. The bioremediation alternative would use a technology that is not proven for this contaminant and would involve additional study to determine effectiveness. Study of the bioremediation process and models indicated that the pesticide may migrate, and the ultimate cleanup effectiveness may not be definable. The three excavation alternatives would remove the pesticide. The calculated safety factors increase and the exposure risks decrease with the increased excavation, hence a cost for exposure risk must be considered. Safety factors rise as the level of excavation effort rises until all of the contaminant is removed.

Remedial Action Selection

A comparison of remedial alternatives suggested that the third excavation alternative most quickly removes the problem and renders the site safe from long-term exposure. Areas of elevated concentrations of pesticide would be excavated and transported to a hazardous waste landfill. The small quantity of pesticide remaining would be covered by an impermeable cap to preclude infiltrating water from leaching the pesticide deeper. The reduced quantity of contaminant and greatly lessened ability for vertical migration produced the greatest benefit for the cost and effort expended. The cost for benefit (safety factor) was the negotiable issue, and the third alternative proved the best compromise to all interested parties.

SUMMARY

Site remediation will be a compromise based on applicable regulations, hydro-geologic data and data interpretation, final cleanup concentrations of contaminants, negotiation of the remedial action plan alternatives, cost, and remediation plan execution. Consultants and contractors cannot completely remove all of the contaminant from the site subsurface. The ability to estimate the safety risk and cost–benefit ratio of the remediation effort should be used. A remedial action plan based on accurate subsurface information, applied within regulatory guidelines and within the ability of current technology and available budget, is the best compromise. Given the myriad legal, scientific, and engineering problems faced in even small-scale cleanups, striving for an unattainable cleanup goal is counterproductive and cost-prohibitive. A maximum effort with the most innovative and efficient use of the appropriate technology, given the site geology and hydrogeology constraints and available funds, will often result in the best remedial approach selection.

REFERENCES

Abdul, A. S., Gibson, T. L., and Rai, D. N., 1990, Laboratory studies of the flow of some organic solvents and their aqueous solutions through bentonite and kaolin clays, *Ground Water*, 28, 524–533.

Aller, L., Bennett, T. W., Hackett, G., Petty, R. J., Lehr, J. H., Sedoris, H., Nielson, D. M., and Denne, J. E., 1989, *Handbook of Suggested Practices for the Design and Installation of Groundwater Monitoring Wells*, U.S. EPA, Las Vegas, NV, in cooperation with the National Water Well Association, Dublin, Ohio; EPA/600/4-89/034, 398 pp.

Aller, L. T., Bennett, T., Lehr, J. H., Petty, R. J., and Hackett, G., 1987, *DRASTIC: a Standardized System for Evaluating Groundwater Pollution Using Hydrogeologic Settings*, U.S. EPA, EPA/600/2-87/035.

American Society of Testing Methods, 1988, *Annual Book of ASTM Standards*, Section 4, Construction, Vol. 04.08, Soil and Rock, Building Stones; Geotextiles; Methods D 420-487, D 653-687, D 2487-2485, D 2488-2484; revised periodically, American Society of Testing Methods, Philadelphia.

Anderson, K. E., 1993, *Ground Water Handbook*, National Groundwater Association, Dublin, Ohio, 401 pp.

Association of Engineering Geologists, 1981, Professional Practices Handbook: Special Publication No. 5.

Barcelona, M. J. and Helfrich, J. A., 1986, Well construction and purging effects on groundwater samples, *Environ. Sci. Technol.*, 20, 11, 1179–1184.

Barker, J. F., Patrick, G. C., and Major, D., 1987, Natural attenuation of aromatic hydrocarbons in a shallow sand aquifer, *Groundwater Monitoring Rev.*, 6, 64–71.

Bartow, G. and Davenport, C., 1995, Pump-and-treat accomplishments: a review of the effectiveness of groundwater remediation in the Santa Clara Valley, California, *Groundwater Monitoring and Remediation*, 15, 140–146.

Bear, J., Tsang, C., and Marsily, G., 1993, *Flow and Contaminant Transport in Fractured Rock*, Academic Press, San Diego, 560 pp.

Behnke, J., Palmer, C. M., Peterson, D., and Peterson, J. L., 1990, *Groundwater Contamination and Field Investigation Methods*; Workshop Notebook, California State University, Chico.

Birkeland, P. W., 1984, *Soils and Geomorphology*, Oxford University Press, New York, 372 pp.

Bjerg, P. L., Rugge, K., Pederson, J. K., and Christensen, T. H., 1995, Distribution of redox-sensitive groundwater quality parameters downgradient of a landfill (Grindsted, Denmark), *Environmental Science and Technology*, 29, 1387–1394.

Blunt, D., Costello, S., and McCardell, B., 1988, Delineation and Remedial Action Planning — a spill of the pesticides 2,4-A and 2,4,5-T, in Proc. of Haymacon 88, Association of Bay Area Governments, Anaheim, 571–585.

Boulton, N. S., 1963, Analysis of data from non-equilibrium pumping tests allowing for delayed yield storage, Proceedings Institutional Civil Engineers, 26, 469–482.

Bouwer, H., 1989, The Bower and Rice Slug Test — An Update, Groundwater, 27, 304–309.

Bouwer, H. and Rice, R. C., 1976, A slug test for determining hydraulic conductivity of unconfined aquifers with completely or partially penetrating wells, Water Resources Research, 12, 3, 423–428.

Brusseau, M. L., 1993, Complex mixtures and groundwater quality: Environmental research brief, U.S. Environmental Protection Agency, EPA/600S-93/004, 15 pp.

Casagrande, A., 1948, Classification and identification of soils, Transactions, American Society of Civil Engineers, 113, 901–930.

Conrad, S. H., Hagan, E. F., and Wilson, J. L., 1987, Why are residual saturations of organic liquids different above and below the water table?, National Water Well Association Petroleum Hydrocarbon Conference, Houston, 19 pp.

Cooper, H. H., Jr., Bredehoeft, J. D., and Papadopulas, I. S., 1967, Response of a finite-diameter well to an instantaneous charge of water, Water Resources Research, 3, 1, 263–269.

Cooper, H. S., Jr. and Jacob, C. E., 1946, A generalized method for evaluating formation constants and summarizing well-field history, Transactions American Geophysical Union, 27(4), 526–534.

Creasey, C. L. and Dreiss, S., 1985, Soil water samplers: Do they significantly bias concentrations in water samplers?, in Proc. NWWA Conf. on Characterization and Monitoring of the Vadose (Unsaturated) Zone, Denver, Nov. 1985, 173–181.

Davis, S. H., 1987, What is hydrogeology?, Groundwater, 25, 2–3.

Davis, S. N. and DeWeist, R. J. M., 1966, Hydrogeology, John Wiley & Sons, New York, 463 pp.

DeRuiter, J., 1982, The static cone penetrations test, state of the art report, in Proc. 2nd European Symposium on Penetration Testing, Amsterdam, Vol. 2, 389–405.

Domenico, P. A. and Schwartz, F. W., 1990, Physical and Chemical Hydrogeology, John Wiley & Sons, New York, 824 pp.

Dragun, J., 1988, The Soil Chemistry of Hazardous Waste, Hazardous Materials Control Research Institute, Silver Spring, MD, 458 pp.

Driscoll, F. G., 1986, Groundwater and Wells, 2nd ed., Johnson Filtration Systems, H.M. Smyth Co., St. Paul, MN, 1089 pp.

Dunlap, L. E., 1985, Sampling for trace level dissolved hydrocarbons from recovery wells rather than observation wells, Proc. Petroleum Hydrocarbons and Organic Chemicals in Ground Water — Prevention, Detection, and Restoration, 223–235.

Dupuit, J., 1863, Etudies theroriques et pratiques sur le mouvement des eaux dans les canaux decouverts et a travers les terrains permeables: 2eme ed., Dunot, Paris, 304 pp.

Elliott, J., The Toxics Program Matrix 1988–1995 (California, Florida, Illinois, New Jersey, Ohio, Pennsylvania, Texas) with yearly updates, Specialty Publishers, Toronto, Canada.

Everett, L. G., Wilson, L. G., and Hoylman, E. W., 1984, Vadose Zone Monitoring for Hazardous Waste Sites, Noyes Data Corp., Park Ridge, NJ, 360 pp.

Ferris, J. G. and Knowles, D. B., 1954, The slug-injection test for estimating the coefficient of transmissivity of an aquifer, U.S. Geological Survey Water Supply Paper 1536-J.

Fetter, C. W., Jr., 1988, Applied Hydrology, C.E. Merrill, New York, 488 pp.

Freeze, R. A. and Cherry, J. A., 1979, Groundwater, Prentice-Hall, Englewood Cliffs, NJ, 604 pp.

Gear, B. B. and Connelley, J. P., 1985, Guidelines for monitoring well installation, in *5th National Symposium and Exposition on Aquifer Restoration and Groundwater Monitoring*, 83–104.

Gibs, J. and Imbrigiotta, T. E., 1990, Well-purging criteria for sampling purgeable organic compounds, *Groundwater*, 28, 68–78.

Gierke, J. S., Hutzler, N. J., and Crittenden, J. C., 1985, Modeling the movement of volatile organic chemicals in the unsaturated zone, in *Proc. NWWA Conf. on Characterization and Monitoring of the Vadose (Unsaturated) Zone*, No. 19–21, 352–371.

Gillham, R. W. and Cherry, J. A., 1982, *Contaminant Migration in Saturated Unconsolidated Geologic Deposits*, Geological Society of America Special Paper 189, 31–62.

Gillham, R. W., Baker, M. J. L., Barker, J. F., and Cherry, J. A., 1983, *Groundwater Monitoring and Bias*, American Petroleum Institute Publication 4367, 206 pp.

Glass, R., Steenhur, E., and Parlange, J., 1988, Wetting front instability as a rapid and far-reaching hydrologic process in the vadose zone, Journal of Contaminant Hydrology, 3, 207–226.

Gustafson, G. and Krasny, J., 1994, Crystalline rock aquifers: Their occurrence, use and importance, *Applied Hydrogeology*, 2, 64–75.

Gymer, R. G., 1973, *Chemistry: An Ecological Approach*, Harper and Row, New York, 801 pp.

Hackett, G., 1987, Drilling and constructing monitoring wells with hollowstem augers: Part I — Drilling considerations, *Groundwater Monitoring Review*, 7, 51–62.

Hantush, M. S., 1956, Analysis of data from pumping tests in leaky aquifers, American Geophysical Union Transactions, 37, 702–714.

Hantush, M. S., 1960, Modification of the theory of leaky aquifers, *Journal of Geophysical Research*, 65, 3713–3725.

Hantush, M. S., 1962, Aquifer tests on partially penetrating wells, American Society of Civil Engineers Transactions, 127(1), 284–308.

Harrill, J. R., 1970, Determining transmissivity from water-level recovery of a step-drawdown test, U.S. Geological Survey Professional Paper 700-C, C212–C213.

Hatheway, A. W., 1994, Computer modeling, Part 1; Accept its use and value with reasoned skepticism, in *Association of Engineering Geologists News*, 37 (winter), 31–34.

Healy, B., 1989, Monitoring well installation misconceptions about mud rotary drilling, in *National Drilling Buyers Guide*, U.S. Govt. Printing Office, Bonifay, FL, 24 pp.

Heath, R. C., 1982, Basic Groundwater Hydrology, U.S. Geological Survey Water Supply Paper 2220, 85 pp.

Heath, R. C., 1984, Ground-water Regions of the United States, U.S. Geological Survey Water Supply Paper 2242, 78 pp.

Heath, R. C. and Trainer, F. W., 1981, *Introduction to Groundwater Hydrology*, Water Well Journal Publishing, 285 pp.

Hem, J. D., 1985, Study and Interpretation of the Chemical Characteristics of Natural Water, U.S. Geological Supply Paper 2254, 3rd ed., USGPO, 263 pp.

Hern, S. C. and Melancon, S. M., 1986, *Vadose Zone Modeling for Organic Pollutants*, Lewis Publishers, Chelsea, MI, 295 pp.

Hillel, D., 1980, *Fundamentals of Soil Physics*, Academic Press, New York, 413 pp.

Hitchon, B. and Bachu, S., Eds., 1988, Proc. 4th Canadian/American Conference on Hydrogeology, National Water Well Association, Dublin, Ohio, 283 pp.

Hodapp, D., Sagebiel, J., and Teeter, S., 1989, *Introduction to Organic Chemistry for Hazardous Materials Management*, University of California, Davis.

Jacob, C. E., 1963, Correction of Drawdowns Caused by a Pumped Well Tapping Less than a Full Thickness of an Aquifer, U.S. Geological Survey Water Supply Paper 1536-I, 272–282.

Jenkins, D. N. and Prentice, J. K., 1982, Theory of aquifer test analysis in fractured rocks under linear (nonradial) flow conditions, Ground Water, 20, 1–21.

Ji, W., Dahmani, A., Ahfeld, A., Lin, J., and Hill, E., 1994, Laboratory study of air sparging: Air flow visualization, Groundwater Monitoring and Remediation, 13, 115–126.

Johnson, R. L., Johnson, P. C., McWhorter, D. B., Hinchee, R. E., and Goodman, I., 1994, An overview of in situ air sparging, Groundwater Monitoring and Remediation, 13, 127–135.

Johnson, P. C., Kembloski, M. W., and Colthart, J. P., 1990a, Quantitative analysis for the cleanup of hydrocarbon contaminated soils by in-situ soil venting, Ground Water, 28, 403–412.

Johnson, P. C., Stanley, C. C., Kembloski, D. L., Byers, D. L., and Colthart, J. P., 1990b, A practical approach to the design, operation and monitoring of in situ soil-venting systems, Groundwater Monitoring Review, 10, 159–178.

Jopling, A. V. and McDonald, B. C., 1975, Glaciofluvial and Glaciolacustrine Sedimentation, Society of Economists, Paleontologists and Mineralogists Special Pub. No. 23, Tulsa, OK, 320 pp.

Jury, W. A., Spencer, W. F., and Farmer, W. J., 1983, Use of model for predicting relative volatility, persistence, and mobility of pesticides and other trace organics in soil systems, in Hazardous Assessment of Chemicals, Vol. 2, Academic Press, New York.

Keely, J. F., 1982, Chemical time series sampling, Groundwater Monitoring Review, 2, 29–38.

Keely, J. F., 1984, Optimizing pumping strategies for contaminant sites and remedial actions, Groundwater Monitoring Review, 4, 63–74.

Keely, J. F., 1989, Performance evaluations of pump-and-treat remediations, U.S. EPA Groundwater Issue, October 1989, EPA/540/4-89/005, 19 pp.

Keely, J. F. and Boateng, K., 1987, Monitoring well installation, purging and sampling techniques — Part I: Conceptualizations: Groundwater, 25, 300–313.

Keys, W. S. and MacCary, L. M., 1971, Application of Borehole Geophysics to Water Resource Investigations, Tech. of Water Resources Inv., U.S. Geological Survey, Chapter E1, Book 2.

Kostecki, P., Calabrese, E., and Oliver, T., 1995, State Summary of Cleanup Standards, Soil and Groundwater Cleanup, November, 16–54.

Kruseman, G. P. and DeRidder, N. A., 1990, Analysis and Evaluation of Pumping Test Data, ILRI-ISBN 90-70260-808, Bulletin No. 11, International Institute for Land Reclamation and Improvement, Wageningen, The Netherlands, 200 pp.

Kueper, B. H., Redman, D., Starr, R. C., Reitsma, S., and Mah, M., 1993, A field experiment to study the behavior of tetrachloroethlyene below the water table: Spacial distribution of residual and pooled DNAPL, Ground Water, 31, pp. 756–766.

Lehr, J. H., 1991, Granular-activated carbon (GAC): Everyone knows it, few understand it, Groundwater Monitoring and Remediation, 10, 5–7.

LeRoy, L. W., LeRoy, D. O., and Raese, J. W., 1977, Subsurface Geology, Petroleum, Mining, Construction, Colorado School of Mines, Golden, CO, 941 pp.

Lewis, R. J., 1993, Revision of Hawley's Condensed Chemical Dictionary, 12th ed., Van Nostrand Reinhold, New York, 1275 pp.

Lindsay, W. W., Chemical Equilibrium in Soils, John Wiley & Sons, New York, 350 pp.

Lohman, S. W., 1979, Groundwater Hydraulics, U.S. Geological Survey Professional Paper 708, U.S. Government Printing Office, Washington, D.C., 68 pp.

MacKay, D. M., Roberts, P. U., and Cherry, J. A., 1985, Transport of organic contaminants in groundwater, *Environmental Science Technology*, 19, 1–9.

Marbury, R. E. and Brazie, M. E., 1988, Groundwater Monitoring in Tight Formations, in Proc. 2nd Annual Outdoor Action Conference on Aquifer Restoration Groundwater Monitoring and Geophysical Methods, Vol. I, Las Vegas, NV, 483–492 pp.

Maslia M. L. and Randolph, R. B., 1990, Methods and computer program documentation for determining anisotropic transmissivity tensor components of two-dimensional groundwater flow: U.S. Geological Survey Water Supply Paper 2308, 1–16.

Matthess, G., 1982, *The Properties of Groundwater*, John Wiley & Sons, New York, 406 pp.

Mathewson, C. W., 1979, *Engineering Geology*, Charles E. Merrill, Columbus, OH, 450 pp.

Mathewson, C. C., 1981, *Engineering Geology*, Bell & Howell Co., Columbus, OH, 410 pp.

McAlony, T. A. and Barker, J. F., 1987, Volatilization losses of organics during ground water sampling from low permeability materials, *Groundwater Monitoring Review*, 7, 63–68.

McCray, K. B., 1986, Results of survey of monitoring well practices among groundwater professionals, *Groundwater Monitoring Review*, 6, 37–38.

McCray, K. B., 1988, Contractors optimistic about monitoring business, *Water Well Journal, NWWA*, May, 45–47.

Merry, W. and Palmer, C. M., 1985, Installation and performance of a vadose monitoring system, in *Conf. on Monitoring the Unsaturated (Vadose) Zone*, National Water Well Association, Denver, 107–125.

Miller, D. W., Ed., 1980, *Waste Disposal Effects on Ground Water*, Premier Press, Berkeley, CA, 511–512.

Montgomery, J. H. and Welkom, L. M., 1989, *Groundwater Chemicals Desk Reference*, Lewis Publishers, Chelsea, MI, 640 pp.

Moore, J. W. and Ramamoorthy, S., *Organic Chemicals in Natural Waters, Applied Monitoring and Impact Assessment*, Springer-Verlag, New York, 289 pp.

Morris, D. A. and Johnson, A. I., 1967, Summary of hydrologic and physical properties or rock and soil materials, as analyzed by the Hydrologic Laboratory of the U.S. Geological Survey 1948-1960, U.S. Geological Survey Water Supply Paper 1839-D, 42 pp.

Morrison, R. D., 1989, Uncertainties associated with the transport and sampling of contaminants in the vadose zone, paper presented at Association of Engineering Geologists Meeting, Sacramento, CA, March, 12 pp.

Neretnieks, I., 1993, Solute transport in fractured rock — Applications to radionuclide waste repositories, in *Flow and Contaminant Transport in Fractured Rock*, Academic Press, San Diego, 39–128.

Neuman, S. P., 1972, Theory of flow in unconfined aquifers considering delayed response of the water table, *Water Resources Research*, 8, 1031–1045.

Neuman, S. P., 1975, Analysis of pumping test data from anisotropic unconfined aquifers considering delayed gravity response, *Water Resources Research*, 11(2), 329–342.

Newsom, J. M., 1985, Transport of organic compounds dissolved in ground water, *Groundwater Monitoring Review*, 3, 41–48.

Nielson., D. M. and Johnson, A. I., Eds., 1990, Groundwater and vadose zone monitoring, American Standard Testing Methods Symposium, Albuquerque, American Society for Testing Materials, Ann Arbor, MI, 313 pp.

Nielson, D. M. and Teates G. L., 1985, A comparison of sampling mechanisms available for small-diameter groundwater monitoring wells, in *5th National Symposium and Exposition on Aquifer Restoration and Groundwater Monitoring*, 237–270.

Norris, L. A., 1966, Degradation of 2,4-D and 2,4,5-T in forest litter, *Journal of Forestry*, 64, 475.

Norris, R. B., Hinchee, R. E., Brown, R., Semprini, L., Wilson, J. T., Kampbell, D. H., Reinhard, M., Bouwer, E. J., Borden, R. C., Vogel, E. J., Thomas, J. M., and Ward, C. W., 1993, In-Situ Bioremediation of Groundwater and Geologic Materials: A Review of Technologies, U.S. EPA, EPA/600/R-93/124, 252 pp.

Palmer, C. M. and Elliott, J. F., 1989, Now My Land's Contaminated: Whom Must I Tell and What Must I Do?, in *Proc. Hazard, West, Long Beach, CA*, Tower-Borner Pub., 537–539.

Palmer, C. M., Peterson, J. L., and Behnke, J., 1992, *Principles of Contaminant Hydrogeology*, Lewis Publishers, Chelsea, MI, 211 pp.

Parker, B. L., Gillham, R. W., and Cherry, J. A., 1994, Diffusive disappearence of immiscible-phase organic liquids in fractured geologic media, Ground Water, 32, 805–820.

Parker, L. V., 1994, The effects of groundwater sampling devices on water quality: A literature review, *Groundwater Monitoring and Remediation*, 14, 120–129.

Patrick, R., Ford, E., and Quarles, J., 1987, Federal statutes relevant to the protection of ground water, in *Legal Issues in Groundwater Protection*, American Law Institute — American Bar Association, Philadelphia, 6–45.

Pearsall, K. A. and Eckhardt, D. A. V., 1987, Effects of selected sampling equipment and procedures on the concentrations of Trichloroethylene and related compounds in ground water samples, *Groundwater Monitoring Review*, 6, 64–73.

Peterec, L. and Modesitt, C., 1985, Pumping from multiple wells reduces water production requirements: Recovery of motor vehicle fuels, Long Island, NY, in *Proc. Petroleum Hydrocarbons and Organic Chemicals in Ground Water — Prevention, Detection and Restoration*, 358–373.

Randall, A. D., Francis, R. M., Frimpter, M. H., and Emery, J. M., 1988, Region 19, Northeastern Appalachians, in Back, W., Rosenshein, J. S., and Seaber, P. R., 1988, Hydrogeology: Geological Society of America, The Geology of North America, Vol. 0–2, Geological Society of America, Boulder, CO, 177–188.

Reed, J. E., 1980, Type curves for selected problems of flow to wells in confined aquifers, U.S. Geological Survey Techniques of Water Resources Inv., book 3, chap. B3, 106.

Reineck, H. E. and Singh, I. B., 1986, *Depositional Sedimentary Environments*, Springer-Verlag, Berlin, 551 pp.

Richards, D. B., 1985, Ground-Water Information Manual: Coal Mine Permit Applications: U.S. Department of the Interior, Office of Surface Mining Reclamation and Enforcement, and U.S. Geological Survey, U.S. Government Printing Office, Washington, D.C., 275 pp.

Robbins, G. A. and Gemmell, M. M., 1985, Factors requiring resolution in installing vadose zone monitoring systems, in *5th National Symposium and Exposition on Aquifer Restoration and Groundwater Monitoring*, 184–196.

Robertson, W. D. and Blowes, D. W., 1995, Major ion and trace metal geochemistry of an acidic septic system plume in silt, Ground Water, 33, 275–283.

Rosenshein, J. S., Gonthier, J. B., and Allen, W. B., 1968, Hydrologic characteristics and sustained yield of principal ground-water units, Potowomut-Wickford area, Rhode Island, U.S. Geological Survey Water Supply Paper 1775, 38 pp.

Rugge, J., Bjerg, P. L., and Christensen, T. H., 1995, Distribution of organic compounds from municipal solid waste in the groundwater downgradient of a landfill (Grindsted, Denmark): *Environmental Science and Technology*, 29, 1395–1400.

Santa Clara Valley Water District, 1989, Investigation and remediation at fuel leak tanks — Guidelines for investigation and technical report preparation, San Jose, CA (March 1989 guidance documentation and later revisions).

Sara M. N., 1994, *Standard Handbook for Solid and Hazardous Waste Facility Assessments*, Lewis Publishers, Boca Raton, FL, sections 1–12.

Sax, N. I. and Lewis, R. J., 1987, *Hawley's Condensed Chemical Dictionary*, Van Nostrand Reinhold, New York, 2188 pp.

Schmelling, S. G. and Ross, R. R., 1989, Contaminant transport in fractured media: Models for decision makers (EPA Superfund Issue Paper): U.S. Environmental Protection Agency, EPA/540/4-89/004, 9 pp.

Schmidt, K. P., 1982, How representative are water samples collected from wells?, in Proc. of the 2nd National Symposium on Aquifer Restoration and Groundwater Monitoring, National Water Well Assoc., Columbus, OH, 117–128.

Schneider, W. J., 1970, Hydrologic implications of solid-waste disposal, U.S. Geological Survey Circular 601-F, 10 pp.

Schwille, F., 1988, *Dense Chlorinated Solvents in Porous and Fractured Media*, Lewis Publishers, Chelsea, MI, 146 pp.

Sen, Z., 1986, Aquifer test analysis in fractured rocks with linear flow paths: Ground Water, 24, in *Fracture Flow Anthology*, National Ground Water Association, April 1992, Dublin, Ohio, article 4.

Skoog, D. A. and West, D. M., 1971, *Principals of Instrumental Analysis*, Holt, Rinehart and Winston, New York, 710 pp.

State of California, 1984(1973) Design Manual, Department of Transportation (revised by University of California, Berkeley).

State of California, 1995, Title 23 Waters, Chapter 3, Subchapter 16 — Underground Tank Regulations, articles 1–10.

Stephenson, D. A., Fleming, A. H., and Mickelson, D. M., 1988, Glacial deposits, in Back, W., Rosenshein, J. S., and Seaber, P. R., Eds., *Hydrogeology: Geological Society of America, The Geology of North America*, Geological Society of America, Boulder, CO, Vol. 0–2, 301–314.

Streltsova, T. D., 1974, Drawdown in a compressible unconfined aquifer, *Journal Hydraulics Division*, Proceedings of American Society of Civil Engineers, 100(11), 1601–1616.

Sullivan, C. R., Zinner, R. E., and Hughes, J. P., 1988, The occurence of hydrocarbon on an unconfined aquifer and implications for liquid recovery, in NWWA Conference on Petroleum Hydrocarbons, Houston, 135–155.

Sykes, A. L., McAllister, R. A., and Homolya, 1986, Sorption of organics by monitoring well construction materials, *Groundwater Monitoring Rev.*, 4, 44.

Testa, S. M. and Winegardner, D. L., 1991, *Restoration of Petroleum Contaminated Aquifers*, Lewis Publishers, Chelsea, MI, 269 pp.

Theis, C. V., 1935, The relationship between the lowering of the piezometric surface and the rate and duration on a well using groundwater storage, American Geophysical Union Transactions, 16(2), 519–524.

Todd, D. K., 1980, *Groundwater Hydrology*, 2nd ed., John Wiley & Sons, 535 pp.

Toth, J., 1984, The role of regional gravity flow in the chemical and thermal evolution of groundwater, in *1st Canadian/American Conf. on Hydrogeology*, Banff, Canada, 3–39.

Trussell, R. R. and DeBoer, J. G., 1983, Analytical Techniques for Volatile Organic Chemicals, in *Occurrence and Removal of Volatile Organic Chemicals from Drinking Water*, American Water Works Association, Coop. Research Dept., Denver, CO, 67–86.

University of Missouri, Rolla, 1981, Seminar for Drillers and Exploration Managers: Short Course Note Set, December 14–16, Phoenix.

U. O. P. Johnson, 1975, *Ground Water and Wells*, U. O. P. Johnson, Saint Paul, MN, 440 pp.

U.S. Department of Agriculture, 1979, Field Manual for Research in Agricultural Hydrology, Ag. Handbook No. 224, 547 pp.

U.S. Department of the Interior, 1981, Ground Water Manual — A Water Resources Technical Publication: Water and Power Resources Service, U.S. Government Printing Office, Washington, D.C., 480 pp.

U.S. Department of the Interior, 1985, Ground-Water Information Manual: Coal Mine Permit Applications, Vol. I by David B. Richards, Office of Surface Mining Reclamation and Enforcement.

U.S. Department of the Interior, 1990, Engineering Geology Field Manual: Bureau of Reclamation, Denver Office, Geology Branch, 598 pp.

U.S. Environmental Protection Agency, 1984a, Permit Guidance Document on Unsaturated Zone Monitoring, for Hazardous Land Treatment Units, EPA/530-800-84-016.

U.S. Environmental Protection Agency, 1984b, Soil Properties Classification and Hydraulic Conductivity Testing, EPA SW-925.

U.S. Environmental Protection Agency, 1985a, Practical Guide for Groundwater Sampling, EPA/600/2-85/10.

U.S. Environmental Protection Agency, 1985b, Seminar Publication, September 1985, Protection of Public Water Supplies from Ground-water Contamination: EPA Center for Environmental Research, Cincinnati, EPA/625/4-85/016, 182 pp.

U.S. Environmental Protection Agency, 1986a, EPA RCRA Ground-Water Monitoring Technical Enforcement Guidance Document, 246 pp. (revised September 1992).

U.S. Environmental Protection Agency, 1986b, Test Methods for Evaluating Solid Waste Physical/Chemical Methods, EPA SW-846, Vols. 1A, 1B, and 1C.

U.S. Environmental Protection Agency (Aller, L. T., et al.), 1987a, Drastic: A Standardized System for Evaluating Ground Water Pollution Potential Using Hydrogeologic Settings, EPA-600/2-89–035, 455 pp.

U.S. Environmental Protection Agency, 1987b, Underground Storage Tank Corrective Technologies, Hazardous Waste Engineering Research Laboratory, EPA/625/6-87-015.

U.S. Environmental Protection Agency, 1987c, Handbook: Groundwater, EPA/625/6-87/016, 212 pp.

U.S. Environmental Protection Agency, 1989, *Transport and Fate of Contaminants in the Subsurface*, Center for Environmental Research Information, Cincinnati, and Robert S. Kerr Environmental Laboratory, Ada, OK, 148 pp.

U.S. Environmental Protection Agency, 1991a, Site Characterization for Subsurface Remediation, Seminar Publication, EPA/625/4-91/026, 228 pp.

U.S. Environmental Protection Agency, 1991b, Description and Sampling of Contaminated Soils, a Pocket Guide, EPA/625/12-91/009, 122 pp.

U.S. Environmental Protection Agency, 1992, RCRA Groundwater Monitoring: Draft Technical Guidance, Nov. 1992 revision, EPA/530-R-93-001.

U.S. Environmental Protection Agency, 1993a, Guidance for Evaluating the Technical Impracticability of Groundwater Restoration, Interim Final, September, Directive 9234.2-25, 26 pp.

U.S. Environmental Protection Agency, 1993b, Wellhead Protection: A Guide for Small Communities, Seminar Publication, EPA/625/R-93/002, 144 pp.

U.S. Environmental Protection Agency, 1993c, Norris, R. D., Hirchee, R. E., Brown, R., McCarty, P. L., Semprini, L., Wilson, J. T., Kampbell, D. H., Reinhard, M., Bouwer, E. J., Borden, R. C., Vogel, T. M., Thomas, J. M., and Ward, C. W., In-Situ Bioremediation of Ground Water and Geological Material: A Review of Technologies, EPA/600/R-93/124, 252 pp.

U.S. Environmental Protection Agency, 1994a, How to Evaluate Alternative Cleanup Technologies for Underground Storage Tank Sites: Solid Waste and Emergency Response, Office of Underground Storage Tanks, EPA 510-B-94-003.

U.S. Environmental Protection Agency, 1994b, Superfund Innovative Technology Evaluation Program, Technology Profiles, 7th ed., November, EPA/540/R-94/526 (and earlier releases).

U.S. Soil Conservation Service, 1978, Groundwater, Water Resources Publications, National Engineering Handbook, Section 18, Engineering Division, Washington, D.C., sections 1–6.

Vishner, G. S., 1965, Use of vertical profile in environmental reconstruction; *Bulletin of the American Association of Petroleum Geologists,* 49, 49–61.

Walton, W. C., 1962, Selected analytical methods for well and aquifer evaluation, Illinois State Water Survey Bulletin 49.

Walton, W. C., 1970, *Groundwater Resource Evaluation,* McGraw-Hill, New York, 664 pp.

White, W. B., 1988, *Geomorphology and Hydrology of Karst Terrains,* Oxford University Press, Oxford, UK, 464 pp.

Williamson, D. A., undated, *The Unified Rock Classification,* U.S. Forest Service, Williamette National Forest, Eugene, OR, 8 pp.

Wilson, C. G., 1980, Monitoring in the Vadose Zone: a review of technical elements and methods, U.S. EPA, Las Vegas, EPA-600/7-80-134.

Wilson, D. D., 1994, Horizontal wells, *Water Well Journal,* 48, 45–47.

Wilson, D. D., 1995, Alternative technologies require new project skills, *Groundwater Monitoring and Remediation,* 15, 75–77.

Zemo, D. A., Bruya, J. E., and Graf, T. E., 1995, The application of petroleum hydrocarbon fingerprint characterization in site investigation and remediation, *Groundwater Monitoring and Remediation,* 15, 147–155.

Index